Safety
in the
Offshore Petroleum
Industry

Safety in the Offshore Petroleum Industry

The Law and Practice for Management

Brenda Barrett, Brian Hindley and Richard Howells

**Kogan
Page**

Acknowledgements

The authors wish to record their gratitude for the valuable assistance they have been given by many colleagues involved with the offshore petroleum industry.

In particular their thanks are due to Bjorn Hendrik Lund of the Norwegian Ministry of Labour and Local Government for much helpful advice; to Arne Stavland and Olav Boye Sivertsen of the Norwegian Petroleum Directorate who undertook the onerous task of commenting on our manuscript; to the Norwegian Maritime Directorate for information and advice, and to colleagues in the United States Federal Agencies, particularly the Occupational Safety and Health Administration, the United States Coast Guard and the United States Geological Survey.

In Great Britain, we have had assistance from many quarters, notably the International Maritime Organisation, the Department of Energy, the Department of Transport and the Health and Safety Executive.

Lastly, our thanks are due to the British Petroleum Co plc and Elf Aquitaine Norge AS for making it possible for the legal authors to get a better understanding of the industry by visiting installations in the North Sea, and to colleagues in the Atlantic Drilling Company for helpful discussions of the safety problems encountered on mobile drilling units.

While we are indebted to many for their help, responsibility for all errors is ours alone.

First published in 1987 by
Kogan Page Ltd
120 Pentonville Road, London N1 9JN

British Library Cataloguing in Publication Data
Barrett, Brenda
    Safety in the offshore petroleum industry.
    1. Offshore oil industry—Safety
    regulations
    I. Title   II. Hindley, Brian
    III. Howells, Richard
    342.3'772   K1835.033

    ISBN 1-85091-070-7

Typeset by V & M Graphics Ltd, Aylesbury, Bucks
Printed and bound in Great Britain by
Billing & Sons Ltd, Worcester

Contents

Preface

It is a pleasure to be invited to introduce this comprehensive book because it gives me an opportunity to review what I consider to be the successful advancement in the development of safety legislation for the North Sea.

My own involvement with the offshore industry in the North Sea dates back to 1970, an extremely interesting time, during which development of the northern North Sea was at its most active. A completely new type and size of offshore installation was being designed, constructed and placed on the sea bed, in deep water, in a hostile environment and at a considerable distance from shore-based support. The magnitude of the installations, the complexity of the operations and the vast number of personnel involved presented the industry with technical problems not previously encountered. This, combined with the rapid development taking place in the northern sector, caught industry and governments in a positon where answers to all problems were not readily available.

Prior to this the offshore industry relied heavily on oilfield standards developed over the years, which on the whole were aimed at onshore operations. They relied heavily on 'best oilfield practice', company 'in house' practice and upon any government legislation existing at the time. Some of this was applicable to North Sea development, but there were many areas not covered and which had to be addressed as 'new ground'.

This is the situation that industry found itself in in the 1970s. Governments were under pressure to legislate, and operating companies, although not completely opposed to the idea, were not in a position to welcome any restriction of their freedom to carry on with the task of constructing new installations and placing new fields in production.

In the UK the Mineral Workings (Offshore Installations) Act 1971 had been enacted and the offshore oil industry was invited by government to participate in the drafting of various regulations under this Act. As one of the industry representatives involved in the preparation of regulations it was both interesting and educational for both parties, government and industry. The opportunity was presented for full and frank discussion from both sides and I believe it is fair to say that on the whole, the end result represented good legislation for both parties at the time. This was one example in which the idea of consultation between government and industry was proven to be workable and was brought to a successful conclusion.

Development of the offshore industry in the North Sea has in the meantime progressed, and some of the Regulations have required amendment; however, on the whole, most of them have withstood the test of time and remain in force.

The authors of this book have treated the development of safety in the offshore industry in a systematic, precise manner and it is designed to take the reader through the background of such development up to the present day. Offshore safety in the North Sea will remain a major issue throughout the life of the producing fields. Anyone involved in this important issue should benefit from reading this book and by having a better understanding of how and why offshore safety remains such an important issue.

The reader is taken systematically through the current legislative provisions and their application to a safe working environment, safe plant and above all a safe worker. Behind this the underlying purpose is of course to avoid accidents which might cause injury to persons or damage to property. The authors are to be commended for their efforts at compiling such a wealth of useful information into this single reference work.

Joseph Howard
E and P Forum
June 1987

Introduction
The General Problem: Developing Safe Systems Offshore

It is now 30 years since the possibility of exploration for and exploitation of submarine petroleum resources was recognised, and the problems associated with acknowledging national rights to engage in industrial activity on the high seas beyond the limits of national jurisdictions were considered worthy of international consideration. The Geneva Convention of 1958 granted, to nations whose coasts abutted on to continental shelves, the right to exercise jurisdiction on their sectors of the shelf to the extent necessary to develop these resources. From this time forward the technical, economic and social challenges of extracting oil and gas from beneath the sea-beds of the world have almost constantly been matters of national and international interest; and not the least matter of concern has been the health and safety of those employed in the offshore industry.

The immense potential of these resources has been of great significance to the world economy; nations fortunate enough to be blessed with offshore oil have enjoyed a peculiar advantage in the international arena. So valuable are these resources, and so significant are they to the economies of the particular countries who can lay claim to them, that national governments have been quick to control their exploitation. Nations have vested in themselves the property rights in these resources. For revenue purposes they have developed systems to determine who shall be entitled to explore for and exploit them, and to ensure that those who are permitted to engage in drilling for and extraction of oil and gas shall both pay for the right to carry out the activity and pay a tax on the hydrocarbons extracted. Indeed, the first priority of governments as exploration began was to regulate for control of these issues.

Within less than 10 years of the Geneva Convention the need to use the power granted to coastal states to legislate to ensure safe systems of work had been recognised. During the following 20 years (from the 1960s to the present time) legislatures – and, to a lesser extent, courts – have been engaged in making and interpreting laws to achieve the safe systems needed to minimise the human cost of exploring for and exploiting submarine petroleum resources.

There has always been a certain glamour about the oil industry. Drilling for oil attracted a mythology of its own: that of tough men fighting the elements and each other for a valuable prize. There was a ruthlessly entrepreneurial aspect to this mythology. Oil people therefore had the reputation of being people who would not only accept the challenges inherent in exploration, but would be prepared to take substantial risks for the sake of substantial gains.

The early concentration of exploration activity in the North Sea brought the oil industry to the shores of a highly developed, heavily industrialised continent, whose governments were well familiar onshore with the techniques of regulatory control for the purposes of both occupational and public safety. These governments were also well versed in legislating and providing for the safety of those who worked on the high seas. The governments of Europe, while anxious to enjoy the economic advantages of an offshore hydrocarbon industry, were reluctant that this industry should be developed if it were likely to be at a high cost in human lives. Therefore they drew on their traditions of legislating for occupational safety onshore and on the high seas, both for standard setting and for regulatory techniques which might be used to control the hazards of the new offshore industry.

The task of legislating for, and subsequently policing of, safe systems offshore has been further complicated because – in a way which is reminiscent of the history of 19th century industrial development onshore – the need for regulatory controls has developed with, but slightly behind, the technical development within the industry. Many aspects of the technology are peculiar to the offshore industry itself, although many of the human hazards – dangerous machinery, lifting operations, etc – are ones with which any onshore factory inspector is very familiar.

It is necessary to find acceptable standards in matters peculiar to the offshore industry, but expertise is obviously confined to a fairly small number of people. The work of the government inspectorates and that of the offshore operators is therefore closely related; the situation lends itself to an interchange of personnel between government enforcement agencies and offshore management of the industry. Debate continues between industry and inspectorates as to whether the problems are peculiar to the particular industry, or whether they are problems inherent in any industrial activity.

The development of the European offshore oil industry has been a story of conflict: conflict with the hostile elements in the North Sea, grappling with technical problems of a scale and complexity hitherto unknown involved in developing the industry in this environment; and conflict between the interests of states in developing these valuable economic resources, while also protecting the lives and providing for the safety, health and welfare of those employed in the industry, with its inherent risks of isolated environment and climatic hazards.

These conflicts have been fuelled by the status and reputation of the international corporations – many of them, initially at least, relying on personnel of North American origin – who have been in the vanguard of the European offshore industry. The differences of emphasis between the work cultures of North America and Europe met in the oil fields of the North Sea; because the relevant oil fields were placed in the jurisdiction of the European states, Europe held itself entitled to require the oil companies to conform to European employment norms. Given that it was the perceived wisdom of the 1970s that the European petroleum industry was more safety conscious than the American, it is interesting to note that a recent report suggests the reverse. However it is not so much the actual, as the perceived, risks which lead to legislative control.[1]

The political, economic and social implications of the development of the offshore industry in the North Sea have been well documented [2] and it is not the purpose of the present authors to go over this ground. This book is concerned only with the regulatory control for safety purposes of the industry, through national parliaments and courts, and within this broad subject area will focus on occupational health and safety legislation.

The theme will be the identification of the balance between governments and industry in the establishment and maintenance of safe systems of work. It will concern legislative provisions for achieving a safer environment, safer plant and safer people. The purpose of seeking such systems is the avoidance of accidents which might cause injury to persons or damage to property. Nevertheless, the possibility that a catastrophe may occur is always present and any attempt to establish safe systems offshore must inevitably pay attention to the full and adequate provision for procedures to minimise loss should emergencies occur. The text traces the stages of offshore petroleum activity from exploration through to removal of the installation, marking the legislative control of the various phases of offshore operations.

Two of the authors are academics but, notwithstanding this, the book is intended for practitioners rather than academics. It concentrates on British and Norwegian laws, partly because these are the regimes with which the authors are most familiar but more essentially because, of the nations whose laws are either written in English or available in English translations, these are two countries who have not only developed comprehensive legal regimes for the control of offshore activities but have done so in an awareness that their physical proximity and the need to co-operate in cross-boundary activity invites comparison between their regulatory systems and enforcement practices. In spite of the close liaison between Britain and Norway in developing their respective offshore interests, the two systems provide interesting comparisons and contrasts to illustrate the variety of tools and techniques available in attaining this control.

The authors have drawn heavily on the laws of these two nations, often quoting from the texts verbatim; relying for this purpose on the unofficial translation into English which is incorporated as part of the published text of the Norwegian legislation. They acknowledge that there are dangers inherent in this practice, in that legal concepts may lose their nuances in translation, but the alternatives of using the Norwegian text or paraphrasing the text in the authors' own words seemed likely to create more problems than they might resolve.

It must be emphasised at this stage that the comparison is limited to the petroleum and related laws of the two countries, removed from their respective legal systems. This constraint is necessary in a work of this nature, which concentrates on the practical issues of importance to those involved in offshore petroleum activity. Otherwise a consideration of, for example, the powers and enforcement techniques available to the respective enforcement agencies in the two countries would involve a consideration of their constitutional, administrative, and criminal law systems, extending into areas far outside the objectives of this work.

An immediate problem which has had to be resolved is that of the choice of terminology: in a comparative study, careful distinctions have to be made between those situations in which words are interchangeable, with national culture leading to the selection of one word rather than another, and those in which the difference of use indicates a difference of meaning. For example, it is well known that Americans say 'gasoline' where the British would say 'petrol'. The authors have noted that the British Department of Energy usually describes the industry here under review as the 'hydrocarbons industry', while it is popularly known in Britain as the 'oil industry'. Norwegian legislation tends to refer to 'petroleum activities'. The authors have decided to use the Norwegian terminology in this instance, as being a convenient, if not entirely accurate way of describing the subject of their enquiry.

Since the book is mainly concerned with legal issues the authors have endeavoured to employ precise legal terminology. This is never an easy task, and it is particularly difficult when comparing two or more regimes. As has already been suggested, there can be difficulties in translation, and legal concepts translate even less easily than popular speech from one language or one regime to another. The task is rendered more difficult when the legislatures have not used clear definitions. For example, while this book is concerned with the legal control of activities on or in the vicinity of offshore installations, there is not strictly any legal definition of 'installation'[3] in either British or Norwegian law: the legislatures have been less concerned with describing the place from which exploration and exploitation may take place and more concerned with describing the activities concerned. In legal systems duties are placed on persons who carry out activities, particularly enterprises. Thus we find duties are imposed in this context on licensees, concession owners, installation owners and senior managers. Throughout this book the authors will endeavour to describe accurately the persons upon whom particular duties are imposed.

The legislation which has been enacted by Parliament in London is described throughout as 'British', since at present the area of major development is in the North Sea off the coast of Great Britain: otherwise it is accepted that when generalising it is a difficult decision whether to use the description 'Britain' or 'UK' – readers should refer to each statute for a precise statement of its application.

The object of the book is to identify and analyse the legal problems with which the offshore operator is likely to have to contend in whatever part of the world he is operating. It is hoped that, even if he is operating in regions where neither British nor Norwegian laws are applicable, consideration of these national laws may give him an indication of likely national approaches. Moreover, in many parts of the world there may be no domestic regulatory legislation to control his activities. In these instances the operator will be guided only by a combination of his humanitarian, moral and commercial instincts, apart from any 'flag' legislation which may be applicable to the mobile installations or shipping he is using. Obviously, no operator will wish to disregard the safety of those who work for him or endanger the very expensive capital equipment with which he is operating. In such cases it may be that this book could give him some guidance as to the sort of hazards to which he should be alerted and the sort of systems he should employ to control them.

Although original texts are quoted extensively, it is not possible to set out the full texts of all the relevant laws, even of the two countries with which the book is most closely concerned. Unfortunately, at the time of going to print the Norwegian situation is complicated by the enforcement agencies' response to the new delegation of powers under the legislation of 1985. Formerly, Regulations emanating from the Maritime Directorate were employed to control both Norwegian mobile installations world-wide and Norwegian and foreign mobiles operating on the Norwegian continental shelf. Today, while the technical standards applicable may remain the same for both shelf and flag activities, the Maritime Directorate's 1986 edition of their *Red Book* gives many of the former Regulations in a re-enacted format relating only to Norwegian flag mobiles. On the Norwegian continental shelf, the Petroleum Directorate continues to enforce the former Regulations relating both to Norwegian and foreign mobiles, as set out in an appendix to their Volume 2 (reproduced in Chapter 6 of this book). In this book it has been necessary on some occasions to give a double reference to shelf and flag versions of what are effectively the same Regulations relating to mobile installations.

Each chapter of the book is headed by a list of the principal statutory materials currently in force referred to within that chapter. Appendix 1 provides a comprehensive list of relevant statutory materials with their reference numbers.

Since legal texts are not very readily available it may be helpful to list here the principal sources of materials and state where they may be obtained:

British: all statutory materials are available either through Her Majesty's Stationery Office or the relevant government departments or organisations concerned, principally the Departments of Energy and Transport and the Health and Safety Executive.

Norwegian: Acts, Regulations and Provisions for the Petroleum Industry. Volumes 1 and 2 are available from Oljedirektoratet, Stavanger.

The *Red Book* (1986 revision) is published on behalf of the Maritime Directorate and is available direct from the publisher, Fabritius Forlag, Oslo.

All Norwegian laws and official publications are available from Lovdals in Oslo.

References

1 *Statistics on Accidents to Offshore Structures Engaged in Oil and Gas Activities in the Period 1970–1985* Part of an annual review by World Wide Offshore Accident Databank. Produced by A/S Veritas, PO Box 300,1322, Hovik, Norway.
2 See, for example, *Labour Law and Offshore Oil* by Kitchen, Jonathan, Croom Helm, 1977; *The Other Price of Britain's Oil* Carson, WG, Martin Robertson and Co, 1982; Routine deaths; fatal accidents in the oil industry, Wright C *The Sociological Review* May 1986.

3 The concept *installation* is not so central to Norwegian legislation as it is to British, but it does nevertheless occur in many Norwegian requirements, eg Petroleum Act 1985, Ss 1, 4 and 24 and the Royal Decree concerning safety of 28 June 1985.

Part I

The Offshore Petroleum Industry
and
The Legislative Framework

1

The Petroleum Industry in the Offshore Environment

This book is concerned with the development of legal regimes for the control by coastal states of exploration for and exploitation of petroleum resources on continental shelves. The emphasis is on the development of protective employment legislation. This chapter is intended to set the scene by providing an outline of how the offshore industry has developed. It will indicate the environmental, technical and human problems which have to be considered and – as far as possible – resolved by the petroleum industry as a prerequisite to successful offshore operation.

Here, as throughout the book, the authors will draw heavily on the British and Norwegian experience in developing the oil and gas resources of the North Sea. The justification for this emphasis is that, although Britain and Norway were by no means the first nations to develop offshore petroleum resources,[1] no other countries have developed their resources with such speed and to the same extent. That they have advanced so rapidly is the more remarkable since there can hardly be another part of the world where the petroleum industry has been developed in spite of the combination of so many environmental, technical, political, economic, social and legal challenges. It may be added, however, that while the focus of discussion is on events which have taken place in the North Sea, the industry which has been in dialogue with the governments of Britain and Norway is an international one.

The petroleum industry, prior to its development offshore, had gained considerable experience of drilling for and extracting oil: this experience had been largely obtained in North America and the Middle East. It is sometimes forgotten that the British government was a major shareholder in a British venture in the Middle East from the early years of the century.[2] Development of the North Sea was much assisted by American onshore experience, and in particular looked initially to American 'good oil field practice' for operational standards.

Another important factor in enabling Britain and Norway to develop their offshore resources was that both countries had a long tradition as seafaring nations, with a familiarity with the maritime environment and developed ports. The affinity of the offshore industry with, and indeed dependence on, traditional merchant shipping activities will be a theme constantly runing through this book.

Development of the North Sea's resources

Development of the North Sea's resources began nearly 30 years ago, when the Geneva Convention[3] recognised that offshore minerals formed part of the natural resources of coastal states which had a continental shelf bordering this or other seas. Considerable interest was and will, into the foreseeable future, be focused on the hydrocarbon deposits which are located below the North and Irish Seas[4] and the Atlantic Ocean bordering the Seaboard of Western Europe, both because of the wealth of these deposits and of their proximity to a rich and highly developed land mass.

Britain was particularly fortunate as it became entitled to a large part of the North Sea, its share being some 46 per cent of the sea-bed below latitude 62°N, with Norway taking some 27 per cent. The search for oil and gas began in earnest in the early 1960s, as the geological surveys and gas find at Groningen in Holland gave clear indications that gas reservoirs were likely to be found, particularly in the southern basin of the North Sea.

The really active search commenced in 1964. The discovery of the West Sole gas field signalled the availability of economic and exploitable quantities of natural gas in the southern basin of the North Sea, below latitude 54°N. Development proceeded apace. The next decade saw seven commercial gas fields developed in this region on the British continental shelf. A further decade, and by 1985 some 14 fields were feeding gas into British terminals at the rate of 43 billion cubic metres (1.5 trillion cubic feet) per annum.[5]

Oil field development was also proceeding fast. With the spectacular Ekofisk discovery of 1969 in the Norwegian sector and the Montrose oil field in British waters, exploration spread swiftly to latitudes north of 56°. By 1977 there were eight British offshore oil fields in production: Argyll, Auk, Beryl, Brent, Claymore, Forties, Montrose and Piper.

Further finds in the Norwegian sector stemmed from the Ekofisk discovery. Thereafter the Bream and Heimdal (condensate) finds of 1972 through to 1974 kept interest alive. The largest find to date, the Statfjord field during 1974, together with Frigg, with the fields extending into the British sector, confirmed Norway as a major offshore producer in Western Europe.

Ten years later, 1985 saw 30 offshore oil fields and 12 offshore gas fields in production on the British continental shelf, with 157 wells drilled during the year. Compare these figures with the Norwegian sector: 10 offshore oil and gas fields in production or under development and 55 wells drilled during the year.

In comparison, the finds of natural gas in the Dutch and Danish sectors of the North Sea have been relatively small. However, their outputs make a significant contribution to the energy needs of both countries.

Technical challenges

The size of the discoveries in the Norwegian and British sectors ensured the future of the offshore industry but by their nature and situation they brought new technical problems for the industry. While offshore drilling was a known technique, its use for the most part had been confined to shallow warm waters by units operating close to the shoreline. The southern basin of the North Sea, with water depths of less than 150 feet, provided conditions which did not strain existing drilling techniques. But there were in addition novel environmental factors where much less was known, relating in particular to weather and sea conditions, current and tidal flows, and sea-bed shifts of sand.

Weather peculiarities were of prime importance as, despite being an important shipping route (which in itself brought its own problems to fixed and mobile installations), very little was known of the problems which would confront and impose restraints on offshore operations. Tides run strongly in southern basin areas, with undercurrents. There are wave and wind patterns to consider and fog is a perennial problem anywhere in the North and Irish Seas and the North Atlantic. Building up the necessary information for the designers and constructors of installations was one thing, holding an accurate weather forecast over a period of 12 hours for those actually working in the area was an entirely different problem. Fog can occur anywhere from between 20 and 35 days, or even more, per year. Severe storms are frequent in more northerly latitudes and in the shallower, more southerly seas gales are all too frequent.

The storms can generate wave heights of over 70 feet and the 100-foot wave has to be considered a possibility. Wind gusts of over 130 knots can be expected. The need for specific and accurate weather forecasts is essential to ensure 'weather windows' for different operations, as well as for giving advance warning of severe conditions which require either the curtailing or closing down of operations for safety reasons. To these factors can be added low mean air temperatures during winter months, allied with mean surface water temperatures north of 56°N of 5–6°C and 6–7°C south of this latitude. This combination of factors can lead to severe icing conditions and the need for de-icing equipment is self-evident.

Today, weather forecast services are provided by commercial organisations and national meteorological services, in part relying on accurate information from selected offshore fields. Additionally, satellite information has a part to play and the computer processing of data is ensuring an improved service with greater accuracy over longer periods.

Shipping and other supply systems

Carrying out petroleum activities at a distance from shore necessitates regular visits of supply vessels to the installations. The original vessels were often unable to operate in winter because of weather conditions. The current generation of supply vessels is more efficient and less likely to be affected by adverse weather. It still remains a hazardous operation, in high winds and rough seas, to position such a vessel alongside an installation in such a manner as to prevent a collision. Dynamically positioned vessels have helped to make the operation more effective and at the same time have ensured safer conditions for those working on the open deck.

The helicopter (the 'chopper') is now the workhorse of offshore transportation. Virtually all personnel in the North and Irish Seas travel to and from their installation by this means. Crew changes by supply boat are the exception; the use of personnel baskets has decreased as Guidance issued in Britain and Regulations in Norway have clearly identified the need to restrict baskets to times of emergency or where it is not practicable to use other means of transport.

From drilling to production

The extent and speed of development caught the oil industry off balance: there was a need for a degree of expedience about the first drilling operations. The southern North Sea, however, with water depths of less than 150 feet and a somewhat milder climate than more northerly latitudes, allowed the jack-up rig to be used. Such rigs are very stable as their legs rest on the bottom.

When under tow, the legs are raised and project only a few metres below the deck. On arrival on station, the legs are lowered by jacks until they rest on the sea-bed with the main deck level some 20 metres above the waves. Their main disadvantage is that they are vulnerable when the legs are being jacked up and down and when moving between locations. However, they are cheaper to operate than other mobile rigs.

Semi-submersible rigs, however, have the great advantage of being able to work in water of much greater depth; up to 200 metres. They vary in size and shape. Most have a rectangular deck, others are pentagon-shaped and the smaller type have triangular decks. They provide a stable platform in rough seas and are supported by pontoon floats submerged some 25 metres below the surface. This means that the supports are not subject to considerable vertical movements, as the effect of waves rapidly diminishes below the surface. Stability is enhanced by an eight-anchor pattern of 15-ton anchors; drilling operations can then continue in all but the roughest conditions.

After the spudding of a well by the drilling rig, there follows a period of analysis to establish whether hydrocarbons are present and whether development is likely to be profitable. It is likely that appraisal wells will be drilled in order to gain more information about the potential reservoirs. If the findings remain promising then production tests are made under conditions as near as possible to those which would exist if the well were in actual production. It is then that a final assessment can be made, followed by an announcement of a commercial find, its potential production rate and potential reserves. Commercial development can then begin – but it may have taken two or more years to reach this stage.

By the time the decision has been made that production will be viable, work will already be advanced on planning what kind of production facilities will be required. Decisions will, for example, have to be made concerning the type and size of installation, the facilities needed, the materials of construction; indeed, whether more than one platform is required. If there will be two or more platforms, it will have to be decided whether they should be permanently bridged. There might even be remotely-controlled satellite platforms.

North Sea production platforms are of concrete or steel, resting on and grouted into the sea-bed. Their size and type are dependent on field conditions, distance from shore, whether there is a pipe-line or the need for tanker loading (necessitating provision for storage). The processing needs will largely determine both the size of the installation itself and the manning requirements; they will also dictate the type and extent of equipment which is to be carried topside, such as gas/oil separators, compressors and pumps.

The structure will also have to carry all the other items of equipment necessary to provide reasonable and tolerable working, sleeping, resting, eating and general welfare conditions for personnel varying in number from less than 20 to over 300. Numbers are swollen during workover periods, and if much topside fabrication is carried out *in situ*, the workforce can perhaps rise to 800 or even 1,000 persons for an initial period of up to 18 months or more.

Like any other operation, when setting up an industrial activity within a limited area it is often necessary to fit everything in almost with a shoe horn. Thus, the equipment, the plant and the machinery is neither easily removed nor replaced when in position. Moreover, in the case of the early installations, certain plant (such as many of the earlier cranes still in use) was not specifically designed for its offshore activity and was adapted for this purpose. Additionally the salt-laden atmosphere can lead to rapid corrosion and deterioration. Thus, there is reason to believe that failure of equipment, plant, etc is relatively high in the areas of drilling, production, maintenance and cranes, particularly the older types. Thorough examinations, with testing where necessary, and adequate and well-implemented maintenance schemes all have a part to play in reducing to a minimum this type of occurrence. The training and skill of the workforce are essential prerequisites. Often – possibly too often – the early warnings of impending failure are not recognised until after the failure has occurred.

Not all the problems, however, are visible. There are considerable difficulties in ensuring that large steel structures will withstand the anticipated sea conditions, the corrosive nature of seawater and the formation of marine growth. Hitherto, concrete rather than steel structures have apparently offered greater overall economies for their owners, but the floating production unit is already effectively operating and must have a future with the smaller marginal fields if position keeping does not, in the light of experience, create too many problems and difficulties.

As production moves to more northerly latitudes, to even deeper waters and increasingly adverse weather conditions, there will be added pressure to move to satellite sea-bed wellheads controlled from the surface from a single installation. The total sea-bed completion, with wellheads, gas/oil separators and pipe-lines on the sea-bed is a feasible operation, remotely or automatically controlled. The possibility of dry transfers of maintenance or even production crews working in one atmosphere is quite feasible. Working in this manner, the problem of acceptable diving working conditions at increased depths and enhanced pressures is at least partly overcome.

Whatever type of installation or subsea completion is used, there must be incorporated into the design, and subsequently into construction, an acceptable and agreed factor of safety. One aspect of this factor is of interest to all who study the offshore oil world. The centre of the Ekofisk field is sinking at a rate which showed a total subsidence of three metres by the end of 1985, as a result of petroleum being extracted from the reservoir three kilometres below the sea-bed. The risk to the platforms serving the field is not yet critical as they have been designed to withstand a total subsidence of six metres. It is intended that a jacking-up operation will elevate the platforms by a further six metres.

However, solutions will have to be found, including modification of design and structure of the platforms so that they can withstand greater subsidence.

Transportation of petroleum

The field will be provided with a pipe-line to shore if this is viable and justified by the size of the reservoir; many well-developed pipe-line complexes have already been laid. Such lines bring their own problems, not least being the need for surface stability during the laying operation despite the lay barge being stabilised with its own ballasting system. The barges operate by attendant vessels laying out an appropriate anchor pattern along which the barge pulls itself by winches during the laying of the actual pipe-line, with the lengths of piping being welded *in situ* on the barge. Reeling out pipe-line up to limited dimensions is now possible, but the barge has to leave station to replenish the pipe reel. Further development has allowed more conventional type vessels to be used to lay pipe-lines up to 24 inches (which can hold station by dynamic positioning). There is a need for meticulous planning as pipe-line laying in the North Sea is restricted for the most part to between April and September.

Raw oil and gas is unsuitable for onward transmission through a pipe-line to the onshore terminal and onwards through the onshore network. In the case of gas; matter, water and liquid hydrocarbons are removed. The residual methanol with hydrocarbon condensate is then propelled along the pipe-line to the shore by neoprene spheres which are introduced for this purpose.

Employment offshore

When the installation is complete and technical problems overcome, to enable the operation of the installation and the transportation of petroleum products from the installation to shore, there still remains a further ingredient to be considered and evaluated: namely, the workforce. The offshore workforce is an ever-changing population: even on the installation itself, only one-third or even one-fifth of the workforce is likely to be directly employed by the installation owner.

Whatever the employment relationships, persons work offshore at least 12 hours a day often for 14 consecutive days before they are relieved. Thus the 'hardware' cannot stand alone: because people are placed on to and into the structure of the installation, the needs of the 'crew' must be considered at the design stage and built into the operational system.

The role and the needs of the crew have to be evaluated, as must the systems adopted which will ensure their safety within the overall plan. The administrative systems must ensure that the parts are knitted together. The calibre of management must be considered and only appropriate personnel appointed. Supervision must be adequate. The level of skill and training of all personnel must be of the highest order and all must be able to work together as a team, both during normal operations and in emergencies and under stress.

Workers engaged on pipe-laying barges often do not set foot on land from beginning to completion of the contract: this may mean they are offshore for months at a time. Such workers may have been brought together from all over the world for just that one contract or for a series of separate operations.

This is a workforce which has to be skilled, fit, accommodated and fed. Its safety, health and welfare must be catered for throughout each 24-hour period. It works in an alien environment: for the most part these workers are not sailors, but construction, factory and service workers who happen to carry out their activities in a maritime environment. Because of this environment they must be self-sufficient, self-motivated and possess the skills and abilities to ensure a safe and healthy environment within the umbrella of the working conditions provided by the licensee, the installation owner, the barge owner and the contractors. The designers, the contractors, the certifying authorities and the governments of the states involved all have a part to play in ensuring that structures and equipment are capable of carrying out the works for which they were designed and under the conditions in which they were expected to operate.

The operators have made provision at least to the regulatory standards, so that cabin allocation ensures at the most four persons per cabin and possibly only two to share, often with built-in shower, washing and toilet facilities on more modern installations. On others, such facilities as showers and toilets are close at hand. Large wardrobes are not required, due to the excellent laundry and drying rooms provided. They are augmented by work clothing changing areas which are situated adjacent to the main accommodation area.

The catering facilities are excellent, with an unofficial league table run as to the gastronomic delights available. Such facilities ensure a sitting for at least half the complement. There are recreation areas, and permanent cinemas are the norm on large installations. Still, such creature comforts are often little enough compensation for personnel who have to work long and – particularly on the drill floor – arduous hours. Given the 12-hour shift and an eight-hour sleep pattern, little or no time is left for recreation and welfare. Obesity and boredom are problems which have still not been satisfactorily overcome.

The environment separates the workforce as a community from services that, in the developed world, are regarded as normal. There is no way, for example, of calling out and receiving, often within minutes, the services of the ambulance, fire and police departments. While operating in a marine world personnel are separated from their families and working in a close-knit community, which can often lead to psycho-social stress with all the possibilities and implications of errors and mistakes in their work and actions.

There is a marked difference in the work-permit systems for foreign nationals. There is no control on the British shelf, whereas in Norway a system is rigidly enforced, and there are similar controls offshore of Canada. In the British sector the lack of restrictions has led to a multinational workforce, with the evolution of the contractual system which leaves only a minority of the employees directly employed by the oil companies, but where the influence of the oil majors remains paramount.

Lessons learnt the hard way

It is a disaster which is remembered. Who today would remember the *Titanic* if it had not hit an iceberg? Nothing concentrates the mind as quickly as a disaster, particularly a spectacular one which occurs at the frontiers of technology. There have been, and there will be, structural and equipment failures. There will be human errors and failings which lead to incidents and accidents. The longer structures and equipment remain in position or are used and reused, the greater will be the risk.

The history of offshore petroleum activities is sprinkled with disasters and failures; moreover, Western Europe and the Americas feature high in the league table of such events. From these catastrophes have come knowledge and a better understanding of the regulatory and operational systems needed to ensure safe working conditions offshore. There is a constant need to refresh our minds on these disasters which have had, and are still having, a significant impact on the regulatory regimes under review, if for no other reason than to avoid complacency as to the standards subsequently set and enforced. There is always a continuing need to examine and re-examine safety standards and to upgrade requirements, not least following the lessons learned from such offshore disasters as:

Sea Gem

In 1965 the collapse of the Sea Gem off the Humber estuary, with the loss of 13 men, brutally emphasised the toll that the work can exact. The accident demonstrated the vulnerability of jack-up rigs when being moved from station. It also demonstrated the need for a statutory system for management of the installation, including a clear line of authority. In Britain the accident led to a tightening of legislative controls over the activities of offshore installations.[6]

Ekofisk

Perhaps offshore oil only came to mean anything in the mind of the man in the street with the Ekofisk 'blow-out'. Oil had 'arrived'. It was on Friday, 22 April 1977, that the casing on well 14 on the Bravo installation required inspection. This entailed removal of the 'Christmas Tree' (the huge safety-valve system clamped to the top of the oil well). It was at this point that a blow-out preventer (BOP) should have been fitted. In fact, it was put on – but upside down. When the error was realised, attempts to turn it round came too late. By this time mud was being pushed back on to the platform. In minutes, without well control, a full blow-out developed and 112 men boarded and launched the survival craft.

The result of the blow-out was spectacular: oil rushed out at 3,000 tons a day creating an oil slick which finally covered some 900 square miles of the North

Sea. By the time that the well was capped eight days later, by the legendary team of Red Adair and Boots Hansen, it was estimated that 22,500 tons of oil and 60 million cubic feet of natural gas had been lost.

The need for an adequate and appropriate system of work, to provide adequate organisational and administrative systems, and positively to identify equipment in its proper mode, is an essential prerequisite of any successful operation. The subsequent Commission of Inquiry report identified all these factors.[7]

The need to ensure such systems are operational is emphasised by the location, the remoteness and the climatic conditions under which offshore exploration and exploitation is carried out.

Alexander L Kielland

On 27 March 1980 the Alexander Kielland, a semi-submersible drilling rig operating as an accommodation unit, capsized while working in the Ekofisk field of the Norwegian sector, with the loss of 123 of the 212 persons on board. Although designed as a semi-submersible rig it was operated as a 'flotel' accommodating persons who were working on an adjacent installation; its original capacity of 80 berths had been extended in a series of conversions between delivery in 1976 to a final figure of 348 berths in 1978.

It seems that the capsize was triggered by the failure of a lower member which held in place and braced one of the columns. Five remaining bracing members were unable to withstand the subsequent loading on the column; they failed due to overload and the column broke away. The buoyancy of the main deck structure arrested the capsize at an angle of 30-35°. Although the deck was entirely watertight, the capsize was delayed insufficiently long for more than a fraction of those on board to abandon the unit.[8]

This accident has considerably influenced Norway in the redevelopment of her regulatory legislation. Nevertheless, considerations of ensuring adequate inbuilt buoyancy still exercise the minds of designers, technical experts and government departments with a responsibility for the offshore industry.

Ocean Ranger

In February 1982 another semi-submersible, the Ocean Ranger, this time in use as a drilling rig, sank while on location in Canadian waters. There were 84 persons on board at the time and there were no survivors.

It would seem that the rig had stopped drilling due to bad weather, when one of the porthole windows in the ballast control room was shattered. Although the ingress of water was stemmed, seawater had found its way into the electrical control panel operating the rig ballast system, causing it to malfunction, opening valves intended to remain closed. Thereafter a list could not be counteracted by using the ballasting pumps due to the lack of inlet pressure at the aft end of the pontoons. The list increased and, although there must be an element of speculation, it would appear that further attempts to regain manual control of the relevant valves failed and final capsize occurred.

This disaster demonstrated the need to have a clear hierarchy of control between drilling and ship's crew. It also illustrated the problems which can arise where there are several enforcement agencies concerned with the regulatory control of installations.[9]

Key Biscayne

The Key Biscayne was a jack-up rig which was lost off the coast of Australia in September 1983. At the time the rig had been under tow from the Timor Sea to Freemantle in gale-force winds and heavy seas; the two lines parted and the rig sank. There were no lives lost: 52 persons were rescued, of whom 47 were drilling personnel.

The occurrence raised serious questions about the construction of the rig, its seaworthiness, the towing arrangements and why drilling personnel were on board while the rig was being moved. It also demonstrated the need to have qualified mariners manning a rig while it is being towed.

West Vanguard

On 6 October 1985 in well 6407/6/2 on Haltenbanken there was an uncontrolled gas blow-out on the West Vanguard drilling rig. The gas caught fire and there was an explosion which caused the death of one person.

Each of these occurrences was followed by an official report. There were many recommendations, several relating to the regulatory structure, particularly in relation to the occurrences in Canadian and Australian waters.

The Ocean Ranger report showed that there were many lessons to be considered: from the need for a fail-safe type of ballasting control which should remain operable up to a 15° angle of list, to the difficulties inherent in deploying lifeboats (enclosed survival craft) in severe weather conditions.

Immediate consideration was given in Britain to the training of ballast control operators[10] and a team of experts drew up specific and detailed recommendations in the form of guidance. The International Maritime Organisation (IMO) is now about to adopt this as an international standard. It is also possible to close off portlights at ballast control rooms located in columns or if impracticable, to have them suitably strengthened. Many improvements will have to wait until adequate consideration has been given to all the implications, with resultant changes in standards, systems and techniques.

Yet, despite the lessons learnt, the risk remains ever-present. It is likely that further lessons have still to be learnt and actions applied. Of all actions which are necessary, safety must come first and foremost – incorporated at the design stage and during construction – and subsequently the structure and its equipment must be maintained, and rigorously so.

Accident and incident reporting

In addition to major catastrophes, less spectacular accidents take their toll. The offshore workforce, no matter how good the environment in which they work or rest, will be exposed to the risk of accident. Accidents do occur, but they are collated and reported in national official annual reports in different ways.

The Norwegians have until now only published statistics in respect of accidents on production installations; it is intended that in future mobile installations will be included. The UK statistical information relates to both production and drilling units together with accidents on attendant vessels. There is no common ground in defining a fatality and it is not unexpected that serious accidents receive different treatment regarding type and causation.

Looking at the position internationally the same discrepancies occur. The classification of minor accidents is in itself a small minefield. A suitable form of standardised international statistical reporting, relayed to a central source who could develop an acceptable form of classification, would be a material asset to those charged with taking appropriate action. This internationally standardised approach would enable them to introduce remedial systems, modify plant and equipment, and change working conditions so as to reduce risk and, if possible, to prevent a recurrence of accidents or incidents. Statistics of incidents or dangerous occurrences which have occurred without injury are also essential weapons, when standardised, in the armoury of any professional safety manager in the unending war to cut down losses, where prevention is the aim as cure is only a palliative.

Without the constraints of a standardised system of reporting, it is impossible and indeed invidious to attempt any comparative exercise which is meaningful or even helpful. With accidents a daily occurrence throughout the offshore world, into which categories do they mainly fall? It is clear that overall accidents associated with the spectacular events such as blow-outs and major fires are not the principal category. For the most part, they are the accidents which happen to most of the world's workforce employed in any form of industrial activity: such commonplace events as falls at the same level by tripping or slipping, falls to a lower level, being struck by falling objects, handling and contact with moving bodies and machinery or equipment in motion.

Which types of occupation appear to be at the greatest risk? Here there is sufficient common ground to indicate the general categories of drilling, maintenance and handling loads as part of crane operations.

Of course, the environment creates its own hazards of low temperatures, bone-chilling winds and the ever-present sea. Perhaps, surprisingly, they are not featured as the causes of accidents. Yet any fall into the sea in northern climes brings with it the risk of death by hypothermia, drowning or simply the shock of immersion, even though the person may not be injured prior to, or as a result of, the fall. Thus, there is the real need to ensure rescue immediately, or within minutes, to improve the chances of survival. Should the person be injured and unable to assist himself, the need for an efficient rescue system manned by adequately trained persons is all too apparent.

The problems associated with multi-casualty incidents are clearly indicated when reading the account of the Alexander Kielland disaster. Even in such cases the presence of the appropriate type of vessels for rescue purposes, the use of helicopters with appropriate night vision and rescue winches, together with adequate communication systems, can go a long way in minimising any final casualty list.

There remain the many incidents which occur offshore which, as dangerous occurrences, do not involve personal injury but which have an enormous risk potential. Here, statistical information is still more difficult to obtain. Structural damage or failure due to the weather must be considerably augmented by similar damage caused by attendant/supply vessels in collision with the installation. One has only to consider in this category the many hundreds of daily supply vessel movements throughout the world to realise the potential risks.

Major fires often have minor beginnings and while not common, they occur for many reasons, not least that of welding operations. For the most part controlled, the isolated nature of the operation must place fires – any fire – in a high potential risk category.

Licensing and regulatory control

In the 20 years which have elapsed since the first gas was piped ashore in Britain, the industry has been mainly in the hands of the large oil multinationals, and the control of offshore operations has centred round the licensing system. The initial permission to carry out searches in offshore areas arises from the granting of licences by the particular country concerned. Such licences are the subject of bids during the licensing rounds, not all the blocks or areas being necessarily taken up. They consist of initially a licence to explore, followed, where a field is found to be viable, by agreements based on a production licence. In between there is the permission to drill as a consent based on a programme for each well which is appraised and agreed.

The licence system is common to maritime countries in the Western World, but differs in detail from country to country. The law in these matters is not too diverse in the countries bordering the North and Irish Seas. Essentially it specifies the various payments which are required in return for the exploration and where relevant, production rights and the royalties required.

The coastal states also exercise control over the standards of construction of installations, not only of the structure itself but the equipment and plant which are to be used. In Norway this control is exercised through the Petroleum Directorate, which is akin to the Department of Energy in Britain. In Britain reliance for certifying structural integrity is placed on the six Certifying Authorities appointed by the Department of Energy.[11] However, the development of control of systems of work has changed markedly in Norway, from tight central control, historically with several ministries involved, to radically more power given to a single government agency, namely the Petroleum Directorate. Systems of regulatory control will be a major theme throughout the book and so will not be developed here.

The operating companies

Such companies have developed associations in the countries in which they operate. In this country it is the United Kingdom Offshore Operators' Association (UKOOA) which has developed voluntary safety systems which are applied to all who work offshore, whether directly or contractually employed. In Norway a similar role has been played by the Norwegian operating companies association (NIFO).

Voluntary standards adopted by the companies ensure that personnel cannot go offshore until their medical credentials have been established and adequate and appropriate fire-fighting and survival training has been given. Heliports in Britain show appropriate videos especially relating to helicopter travel. Survival suits are now standard and worn for offshore flights and also for in-field flights of short duration. On arrival at the installation, there is almost always a safety briefing, together with allocation of a survival craft, explanation of the appropriate action to be taken in times of emergency, and familiarisation with the warning and alarm signals in use.

References

1   Eg development in California, the Gulf of Mexico and Venezuela began earlier.
2   The Anglo-Iranian Oil Company.
3   See Chapter 2.
4   At the time of writing the UK has not initiated developments off the coast of N Ireland. The Petroleum Act 1987, S.19, will enable the granting of licences in this region.
5   *Development of the Oil and Gas Resources of the UK 1986* Department of Energy.
6   *Report of the Inquiry into the Causes of the Accident to the Drilling Rig Sea Gem 1967* HMSO, Cmnd 3409.
7   *Report No 65 to the Storting (1977-78): The Uncontrolled Blow-out in the Ekofisk Field on 22 April 1977* Ministry of Local Government and Labour, Oslo.
8   *Report No 67 to the Storting (1981-82): The Alexander L Kielland incident* Ministry of Local Government and Labour, Oslo.
9   *Royal Commission on the Ocean Ranger Marine Disaster on 15 February 1982: Report 1984* St Johns, Canada.
10 There was a working party of experts under the chairmanship of one of the authors (Brian Hindley).
11 American Bureau of Shipping; Bureau Veritas; Det Norske Veritas; Germanisher Lloyd; Lloyds Register of Shipping; Offshore Certification Bureau.

2

The Geneva Convention and the Development of Offshore Jurisdictions

Principal legislative provisions

The Geneva Convention on the Continental Shelf 1958.
  Reports of the Proceedings of the 3rd United Nations Conference on the Law of the Sea (The Jamaica Convention).

British legislation
Petroleum (Production) Act 1934.
Radioactive Substances Act 1960.
Continental Shelf Act 1964.
Mineral Workings (Offshore Installations) Act 1971.
Petroleum and Submarine Pipe-lines Act 1975.
Health and Safety at Work, etc Act 1974.
Oil and Gas (Enterprise) Act 1982.
The Petroleum Act 1987

Norwegian legislation
Royal Decree of 5 April 1963* authorising Maritime Directorate to issue regulations.
Royal Decree of 31 May 1963 asserting Norwegian sovereignty over Norwegian continental shelf for purposes of exploration and exploitation of natural deposits.
Act No 12 of 21 June 1963† permitting exploitation under licence.
Royal Decree of 8 December 1972* providing legal framework for offshore operations.
Royal Decree of 3 October 1975* relating to safe practices in exploration and drilling.
Royal Decree of 9 July 1976* relating to safe practices for production etc.
Act No 4 of 4 February, 1977 relating to worker protection and working environment Act of 22 March 1985 no. 11 relating to petroleum activities.
Royal Decree of 13 September 1985 relating to Worker Protection and Working Environment Act.
Norwegian Penal Code

---

* Now wholly or largely superseded by new framework legislation: Act No 11 of 22 March 1985 on petroleum activities and the following Royal Decrees:
Royal Decree of 14 June 1985 supplementing the Act
Royal Decree of 28 June 1985 giving regulations concerning safety
Royal Decree of 28th June 1985 on the licensee's internal control.

† from 1 July 1985 – Act pertaining to scientific research.

Australian legislation
Navigation Act 1912.
Petroleum (Submerged Lands) Act 1967.
Seas and Submerged Lands Act 1973.
Petroleum (Submerged Lands) Amendment Act 1980.

Canadian legislation
Canada Oil and Gas Drilling Regulations 1980.
Newfoundland and Labrador Petroleum Drilling Regulations 1977.

United States legislation
Outer Continental Shelf Lands Act 1953 as amended.
Occupational Safety and Health Act 1970.

International treaties
UK–Norway Delimitation Agreement 10 March, 1965.
Agreement relating to the Ekofisk Pipeline 1973.
Agreement relating to the Exploitation of the Frigg Field Reservoir 1976.
Agreement relating to Murchison 1979.

The right of a coastal state to the natural resources on its continental shelf was first suggested by President Truman in September 1945. By the early 1950s offshore exploration and production had become an industry in its own right. The development of the present activities associated with the exploration for and the exploitation of petroleum resources outside territorial waters was facilitated by the Conventions which emerged from the First United Nations Conference on the Law of the Sea held at Geneva in 1958.[1] This Convention on the continental shelf provided for the necessary extension of national laws beyond their traditional territorial limits.

The conventions emerging from the Geneva Conference were prompted by, and most immediately concerned with, the desire of the European states to develop the petroleum resources of the continental shelf of the North Sea, but they were potentially of relevance to all the continental shelf areas throughout the world. They are now realised to be of significance not only for exploration and exploitation in the North and Irish Seas and North Atlantic Ocean, but also for such activities off the east coast of North America and in the Antipodes. It is, however, open to question how far a Convention is binding upon nations who are not signatories to it.[2]

Traditional limits to national jurisdiction

The traditional concept of national sovereignty was that the territorial jurisdiction of a state did not extend beyond its own shores, or at most beyond territorial waters surrounding its coast (known as its territorial sea); the exact extent of particular territorial waters being in the last instance a matter for the state concerned. Territorial waters are waters within three miles of the coast in the case of Britain. There is no present significance in the three mile limit; it is just the limit which has traditionally been accepted among nations. It may well be changed.[3]

The laws of a state cannot normally be enforced beyond its shores; even within territorial waters the jurisdiction of the courts is less than complete. General exceptions to the territorial boundaries to jurisdiction are that national law may provide for the trial, within the jurisdiction, of its nationals for crimes committed outside the jurisdiction and that a country can enforce its laws on its registered vessels even in respect of what occurs when that vessel is on the high seas. In the present century both these concepts have been extended to the use of nationally-registered aircraft.

The Geneva Convention

The Convention on the Continental Shelf, which the United Nations Conference on the Law of the Sea at Geneva adopted in 1958, forms the basis upon which the coastal states abutting on the North Sea extended their jurisdiction into the North Sea for the purposes of exploration for and exploitation of hydrocarbon resources on their respective sectors of the continental shelf. This experience provides a pattern for the guidance of developments in other continental shelf areas of the world.

The Convention did not, however, envisage the extension by coastal states of full sovereignty on to the adjoining continental shelf: it envisaged no more than the expansion of national jurisdiction to the extent necessary for the exploration for and exploitation of mineral resources. The Convention did not contemplate the grant to the coastal states of any proprietorial rights over the shelf. It was intended that a coastal state would merely have an exclusive right to the resources in its sector of the shelf and should exercise only such governmental powers as were necessary for the enjoyment of these resources. Apart from this limited jurisdiction of the coastal states, the continental shelf would remain a part of the 'high seas' and Article 5 of the Convention specifically provided that exploration for and exploitation of natural resources must not result in any unjustifiable interference with navigation, fishing or the conservation of the sea, nor result in any interference with fundamental oceanographic or other research.

For the purposes of the Convention, the continental shelf was defined as the area of the territorial sea to a depth of 200 metres, or beyond that depth to the possible limits of exploitation of the minerals.

Article 6 of the Convention provides that where the same continental shelf is adjacent to the territories of two or more states whose coasts are opposite each other, the boundary of the continental shelf appertaining to such states shall be determined by agreement between them. In the absence of such agreement, and unless another boundary line is justified by special circumstances, the boundary shall be the median line every point of which is equidistant from the nearest points of the baselines from which the breadth of the territorial sea of each state is measured.

Subsequent history of the Convention

Britain ratified the Convention on 11 May 1964 and being the 22nd ratification, caused the Convention to come into force 30 days later, on 10 June 1964.

The boundary between the British and Norwegian sectors of the North Sea continental shelf was delimited on the broad application of the equidistance principle in 1965, notwithstanding that the Norwegian Trough gave Britain plausible grounds for arguing that its continental shelf extended beyond the median line to the westward edge of that Trough.

The Convention is, however, binding only upon the signatories to it. The particular configuration of the coast of Germany meant that she could lay claim to only a small area of the continental shelf of the North Sea according to the criteria of Article 6. She therefore abstained from signing the Convention and challenged its authority in the International Court of Justice.[4] The Court confirmed that the Convention was not binding upon states which were not signatory to it and held that, as Article 6 was not part of customary international law, it was not binding on coastal states independently of their agreement to the Convention. The Court found that there was nevertheless an obligation on opposite or adjacent states who were not signatories to the Convention to negotiate to decide an equitable boundary between their jurisdictions.

Norway ratified the Convention in 1971. Britain and Norway have entered into agreements concerning the Ekofisk pipe-line, and for the exploitation of the cross-boundary Frigg and Murchison Fields.[5]

Federal states

In the Antipodes, the problem of exercising the rights granted to coastal states by the Geneva Convention was complicated by a dispute by federal and state governments. The federal state (the Commonwealth) enacted the Petroleum (Submerged Lands) Act 1967, which permitted the assumption that rights in the territorial seas and on the continental shelf were vested in the individual states,[6] and proposed a format for state legislation. Individual states enacted their own legislation based on the federal Act. Section 157 of the federal Act enabled the states to make Regulations for the offshore industry: although none of the regulation-making powers expressly referred to safety, arguably some might be exercised for this objective.

To clarify the position between the Commonwealth and the states, the Commonwealth enacted the Seas and Submerged Lands Act 1973. This declared that sovereignty, 'in the territorial sea, its bed and subsoil is vested in and exercisable by the Crown in the right of the Commonwealth'. The states unsuccessfully challenged the proposition that sovereignty was vested in the Commonwealth.[7]

Thus it was established that the Commonwealth could if it wished legislate for the whole offshore field. However, the Petroleum (Submerged Lands) Amendment Act 1980 enabled the creation of joint State–Commonwealth Authorities to regulate the industry.

Similar problems have been encountered on the east coast of North America. In this case, the situation is further complicated because there are two federal governments – Canada and the USA – and each has its member states. The problems of competition for jurisdiction between Canada and the state of

Newfoundland was apparent in the official report of the Ocean Ranger disaster.[8]

Sovereign rights under the Convention

It has been noted that the Convention does not purport to grant to a coastal state full proprietorial rights over its sector of the continental shelf. Under Article 2 of the Convention the coastal states may exercise sovereign rights over the continental shelf only to the extent necessary for exploring it and exploiting its natural resources. Thus a state may not exercise full governmental powers, but only such powers as the Convention permits and as are necessary for the enjoyment of its exclusive rights to the natural resources on the shelf.

Under Article 5 the coastal state is entitled to construct and maintain or operate installations and other devices necessary for the exploration and exploitation of its natural resources, and to establish safety zones around such installations and devices and to take in those zones measures necessary for their protection. The safety zone referred to in the Article may extend to a distance of 500 metres around the installations and other devices which may have been erected, measured from their outer edge. Ships of all nationalities must respect safety zones.

The term 'installation' was not defined for the purposes of the Convention. The Convention does make it clear, however, that installations, although under the jurisdiction of the coastal states, do not acquire the legal status of islands, and must be removed by the coastal state when their purpose is accomplished.

The failure to define 'installation' has been carried forward into national regulatory legislation. In the present context it could have important implications, including determination of the circumstances in which safety zones are permitted by the Convention. On the lack of clarity within the Convention as to what is meant by installation, and the possible implications of this lack of clarity, it has been noted:

> It is clear that production platforms are included, as are such associated devices as the well itself, pipe-lines from the well to the platform, offshore oil storage terminals and single buoy moorings, and excluded are pipe-lines to the shore; but it is not altogether clear from the wording of the Convention if drilling ships or semi-submersible drilling units fall within its terms ... What is principally at issue is whether a drill ship or a semi-submersible drilling unit is entitled to a safety zone.[9]

However, whatever the analysis may suggest concerning safety zones, in practice coastal states have shown no hesitation in legislating petroleum laws for the control of both fixed and mobile structures engaged in petroleum activities. Such restrictions as Article 5 imposes have been taken to relate to installations when moving to and from the location where they are to be stationed.

Article 4 of the Convention supports the freedom of the high seas guaranteed in Article 2 of the Geneva Convention on the High Seas of 1958[10] by stipulating that the coastal state may not impede the laying or maintenance of submarine cables or pipe-lines on the continental shelf. Article 2 gives the coastal state equal freedom with other states to lay pipe-lines anywhere on the continental shelf, including within its own sector.

Apart from the provisions of the Convention, as subsequently interpreted and extended, the coastal state has no internationally recognised rights to exercise jurisdiction on the continental shelf, except for such rights as it may exercise over ships registered under its flag.

Third UN Conference

There have been no material changes in international law relating to offshore petroleum activities since the Geneva Convention, although the issues dealt with by the Convention have been kept under review. In the past decade, the Third United Nations Conference on the Law of the Sea has met from time to time and in 1982[10a] formulated a new international convention on the development of the mineral resources of the sea-bed. This Law of the Seas Convention was signed at Montego Bay, Jamaica on 10 December 1982 by 117 states, including Norway. It will enter into force as soon as 60 countries have ratified or acceded to it. Currently it has 32 signatures. Britain and the USA have yet to sign. The proposals include the extension of national entitlement to allow development by nations who cannot claim coastal state status. Article 60 deals with removal of installations and with safety zones. However, these proposals have still not been generally accepted and although the Convention has now been signed by a number of countries it has yet to be ratified. Nevertheless it has been suggested:

> It seems likely that state practice will conform or come to accept many of the particular convention rules as being binding as custom.[11]

Development of British law

The Continental Shelf Act 1964

The Continental Shelf Act 1964 was passed at the time Britain ratified the Geneva Convention, in order to give Britain jurisdiction over its sector of the continental shelf in the North Sea to the extent permitted by the Convention. Thus it vested in the Crown: 'Any rights exercisable by the UK outside territorial waters with respect to the sea-bed and subsoil and their natural resources ...'[12] and extended the onshore licensing power given in the Petroleum (Production) Act 1934 to enable the licensing of activities on the continental shelf of the North Sea. Following the scheme of the Convention, the statute extended territorial jurisdiction only in respect of installations[13] and the areas surrounding them; it did not purport to relate such control to the whole of the British sector of the continental shelf.

The framework created by this Act has remained the basis of British offshore jurisdiction even though many of its provisions have been amended by subsequent legislation, particularly the Oil and Gas (Enterprise) Act 1982.

Section 2 of the Continental Shelf Act[14] provided for the declaration of safety zones. Section 3 provided for the application of the criminal and civil law of the UK to installations and safety zones in designated areas. Subsequent sections of the Act referred to the extension to these installations of the whole or part of specific statutes such as the Radioactive Substances Act.[15]

Whether a statutory provision such as S3 is capable of affecting the territorial jurisdiction of the whole of the common law and all other regulatory legislation and whether, if conceptually this is possible, the actual wording employed by the section succeeded in bringing about a wide extension of national law on to the continental shelf of the North Sea, raises complex legal arguments which do not appear to have been fully resolved.[16] It is beyond the ambit of this book to pursue the problem at length but the following points may be noted in passing:

1. If S3 intended to bring about the automatic extension of the whole body of British law offshore it is difficult to understand why subsequent sections of the Act made specific provision for the extension of named pieces of legislation. The uncertainty is illustrated by the Misuse of Drugs Regulations 1985; these purport to regulate the use of drugs on installations although the parent Act, the Misuse of Drugs Act 1971, has not been expressly related to the continental shelf except indirectly under licensing provisions.[17]

2. There is a strong common law presumption against domestic laws having extra-territorial force and there is a rule of statutory interpretation that the legislature does not normally intend to alter the common law. There are a number of judicial statements of high authority which demonstrate the reluctance of the British courts to give British laws extra-territorial effect.[18]

3. Even on the widest possible interpretation of the effect of S3, it is difficult to understand how the section could have the effect of applying offshore statutes that manifest a clearly expressed intention to the contrary.

4. Much of the British regulatory legislation, such as the Factories Act 1961 or the Road Traffic Act 1972, may well be inapplicable offshore simply because the circumstances which the legislation seeks to regulate are not found offshore: it is possible that installations might be deemed factories, but there are no roads on offshore installations. The aims of other legislation (such as the Public Health Act) may be relevant to offshore activities, but the enforcement agencies may not be entitled, or organised, to exercise their powers offshore.

5. Instructions subsequently given to licensees holding exploration and production licences have required the licensees to observe certain legislation, such as the Misuse of Drugs Act. It is not clear why such contractual obligations should have been spelt out if these terms were merely repetitive of pre-existing statutory obligations.

There does not seem to have been any litigation in which (with or without reference to S3 of the Continental Shelf Act) the relevance of domestic law to offshore activities has been argued and determined in any British court.

Moreover, although these matters are now governed by SS22 and 23 of the Oil and Gas (Enterprise) Act 1982, which deal respectively with the criminal and the civil laws, the new provisions do very little to clarify the situation. The new sections appear to intend that national courts shall have jurisdiction only on matters specifically provided for by Orders in Council. Thus it would appear that neither the body of the common law nor the body of statute law applies to either mobile or fixed installations in the British sector of the continental shelf.[19] Shipping laws may incidentally bring British registered vessels and their crews, operating as mobile installations in the British sector of the shelf, more clearly within the jurisdiction of British courts. It is arguable that, since the Act of 1982, the presumption against the extension offshore of the body of the criminal law relating to offences against the person and homicide is stronger than it was before.

Section 22(2) of the 1982 Act makes provision for the police to have, in relation to installations, all the powers and privileges they enjoy onshore; however these powers and privileges can only be in respect of enforcement of such law as applies offshore.[20] Moreover, the section has yet to be brought into force.

The Continental Shelf Act acknowledges a potential for Britain to exercise control over exploration for and exploitation of petroleum resources in any part of her sector of the continental shelf of the North Sea, but envisages that such control will only be effected in respect of activities taking place in 'designated areas'.[21] Most of the British sector of the North Sea continental shelf has now been designated as the result of a succession of Orders in Council. A further series of Orders in Council has also given the courts of England, Scotland and Northern Ireland jurisdiction over matters occurring in designated areas. The Oil and Gas (Enterprise) Act S22(5) provides that Orders in Council may extend either criminal or civil law to an area which:

(a) is in a foreign sector of the continental shelf; and
(b) comprises any part of a cross-boundary field.

Provision has in fact been made for the application of employment protection legislation beyond the British sector.[22]

British regulatory legislation

There are four principal statutes – some of which have Regulations made under them – which are relevant to exploration for and exploitation of petroleum resources in the British sector of the North Sea. In a wider context the Merchant Shipping Acts are also relevant, but this chapter will consider only that legislation which was either made especially for, or by extension has been given particular relevance to, offshore hydrocarbon activity.[23]

Petroleum (Production) Act 1934

This Act was passed in order to control the petroleum extraction industry onshore within Great Britain. Its primary purposes were to vest the property in petroleum in Great Britain in the Crown and to empower the granting of licences to persons 'to search for and get' petroleum: it empowered Regulations to be made concerning the granting of licences. Section 6(1) required the then Board of Trade to prescribe model clauses for incorporation in licences. This Act was extended to exploration for and exploitation of petroleum resources on the continental shelf by S1 of the Continental Shelf Act 1964. Section 1(4) of that Act required that model clauses should include provision for the safety, health and welfare of persons employed in operations under the authority of any licence granted under the Act.

Mineral Workings (Offshore Installations) Act 1971
This Act was largely inspired by the official report which followed the accident to the drilling rig 'Sea Gem'.[24] It was a response to the actual development of exploration and exploitation activities in the North Sea in the light of an increasing awareness of the magnitude of the risks involved in these activities. It was passed specifically to regulate these activities, through criminal sanctions imposed in the interests of the health, safety and welfare of the persons employed on installations in designated areas of the British sector of the continental shelf and in British territorial waters.

The following Regulations which have been made under the Act and all but one (marked *) are still in force:

The Offshore Installations (Registration) Regulations 1972.
The Offshore Installations (Managers) Regulations 1972.
The Offshore Installations (Logbooks and Registration of Death) Regulations 1972.
The Offshore Installations (Inspectors and Casualties) Regulations 1973.
The Offshore Installations (Construction and Survey) Regulations 1974.
The Offshore Installations (Public Inquiries) Regulations 1974.
The Offshore Installations (Diving Operations) Regulations 1975*.
The Offshore Installations (Application of the Employers' Liability (Compulsory Insurance) Act 1969) Regulations 1975.
The Offshore Installations (Operational Safety, Health and Welfare) Regulations 1976.
The Offshore Installations (Emergency Procedures) Regulations 1976.
The Offshore Installations (Life-saving Appliances) Regulations 1977.
The Offshore Installations (Fire-fighting Equipment) Regulations 1978.
The Offshore Installations (Well Control) Regulations 1980.
The Offshore Installations (Life-saving Appliances and Fire-fighting Equipment) (Amendment) Regulations 1981, 1983 and 1985.
The Offshore Installations (Included Apparatus or Works) Regulations 1982.
The Offshore Installations (Application of SIs) Regulations 1984.

Petroleum and Submarine Pipe-lines Act 1975
A large part of this Act is concerned with matters unrelated to health and safety. No person may construct or use a pipe-line[25] in territorial waters or in any designated area without authorisation and authorisation may be granted subject to conditions. Additionally, Regulations may be made for securing the proper construction and safe operation of pipe-lines, preventing damage to pipe-lines and securing the safety, health and welfare of persons engaged in pipe-line works. Three sets of Regulations have been made:

The Submarine Pipe-lines (Diving Operations) Regulations 1976.
The Submarine Pipe-lines (Inspectors, etc) Regulations 1977.
The Submarine Pipe-lines Safety Regulations 1982.

The diving Regulations have for all practical purposes been superseded by the Diving Operations at Work Regulations made under the Health and Safety at Work, etc Act below.

Health and Safety at Work, etc Act 1974
The broad objectives of this legislation include securing the health, safety and welfare of persons at work and protecting persons other than persons at work against risk to health or safety arising out of, or in connection with, activities of persons at work. The Act imposes wide general duties upon employers, employees, contractors, manufacturers and controllers of premises and gives very wide regulation-making power.

This Act initially applied only to work activities within the UK but the Health and Safety at Work, etc (Application outside Great Britain) Order 1977 extended it to territorial waters and to designated areas in relation to offshore installations, any activities on and certain specified activities in connection with, installations. It applies also to pipe-line works and certain specified activities in connection with pipe-line works.

It will be observed that the Act applies to certain activities which are not regulated by the Mineral Workings Act. The Order is in the process of being amended to ensure that it covers accommodation in the form of flotels and gas storage installations. By no means all the Regulations made under this Act have been extended offshore but there is a tendency for Regulations which have been made under the Act since the Order in Council to be made to include application offshore. Regulations which apply offshore are:

Lead at Work Regulations 1980.
Diving Operations at Work Regulations 1981.
Asbestos (Licensing) Regulations 1983.
Freight Containers (Safety Convention) Regulations 1984.
Asbestos (Prohibition) Regulations 1985.
Ionising Radiations Regulations 1985.

Employment Protection Legislation
Onshore employment protection legislation, such as the provisions protecting employees from unfair dismissal, has in many instances been extended offshore by Orders in Council.

Development of Norwegian law

At the beginning of the 1960s Norway had no experience of exploration for and exploitation of petroleum resources, so with the development of activities in the North Sea it was necessary to formulate a system of regulatory control *de novo*. The Norwegian response to this situation was systematic and thorough: a framework was established between 1962 and 1976 and the second half of the 1970s saw the promulgation of a number of Regulations containing, in some cases, remarkably detailed standards. This whole system has been subjected to close review since the beginning of the 1980s and much of the framework legislation has now been replaced by a new principal Act, No 11 of 22 March 1985, and three Royal Decrees made further to it in June 1985.[26]

Two fundamental changes in approach have been affected by the new regime introduced by the Act of 1985. First, the emphasis is now firmly on the control of *petroleum activities*, thus many of the former distinctions between legislation for the control of mobile and fixed installations will disappear.[27] Second, the emphasis has been placed on the licensee's responsibility to achieve *internal control*; it is for the licensee to ensure compliance with the detailed regulatory standards through his management systems, except where he has been granted a dispensation from particular Regulations. The detailed Regulations are, at the time of writing mainly as originally published (although many of them have been subject to periodic minor review), but it is intended that in due course the Regulations, like the principal legislation, will be considerably revised. The emphasis in future is likely to be on guidelines rather than Regulations to implement the framework legislation.

The overall picture is, for the time being, a somewhat untidy and, to the authors, confusing combination of the old and the new.

Jurisdiction on the continental shelf

On 31 May 1963 a Royal Decree asserted sovereignty in the following words:

> The sea-bed and its subsoil in the submarine areas outside the coast of the Kingdom of Norway are subject to Norwegian sovereignty in respect of the exploitation of and exploration for natural deposits to such extent as the depth of the sea permits the utilisation of natural deposits, irrespective of any other territorial limits at sea, but not beyond the median line in relation to other states.

This Act observed the terms of the Convention in that it did not attempt to assert sovereignty for purposes other than exploration for and exploitation of natural resources; but did not follow it in respect of the geographical limits it adopted. It was also enacted before Norway had reached any agreement with Britain as to the boundary between the two national jurisdictions.

In contrast with Britain, Norway appeared to be asserting sovereign rights over the whole of its sector of the North Sea continental shelf rather than in designated areas, so preserving a wide latitude for the making of Regulations applicable to activities on the continental shelf. This assertion of sovereignty was generally regarded as sufficient to apply the law of Norway in its entirety to the Norwegian sector of the continental shelf of the North Sea, including the Norwegian Penal Code; thus S12 of the latter which related to acts 'in the realm' was taken henceforth to include the continental shelf.

On this basis a great deal of 'onshore' legislation was applied and observed offshore; for example, the Code of Regulations for electrical installations has been applied to fixed installations offshore. Nevertheless, certain doubts were raised by a case which went to the Norwegian Supreme Court in 1974[28] as to whether S12 of the Penal Code had been effectively extended offshore by the Royal Decree. As a result a statute was passed to clarify the matter by specifically amending the Penal Code so as to apply it to activities on the continental shelf.[29] Section 2 of the Petroleum Act 1985 now expressly applies Norwegian law to installations (other than foreign flag mobiles) except where specific provision to the contrary has been made.

Assertion of state control
An Act of 12 June 1963, relating to exploration and exploitation of submarine natural resources, legislated for activities both in territorial waters and on the continental shelf up to the median line in relation to other states. It went further than the Royal Decree of 31 May in that it might be applied to activities outside the Norwegian part of the continental shelf if such application had been specifically agreed with a foreign state or conceded by international law. The purpose of this legislation was to vest the property in submarine natural resources in the state, to prevent exploration and exploitation without state consent, and to authorise the making of Regulations concerning the exploration for and exploitation of submarine natural resources. The provisions of this Act are now substantially repeated in the new framework Act of 22 March 1985, in S67.

Framework Royal Decrees
The Act of 12 June 1963 was followed by a series of Royal Decrees which were for the most part made pursuant to the regulation-making power granted in that Act. These Decrees established a broad framework for control of natural resources on the continental shelf but did not all relate either to safety or to exploration for and exploitation of petroleum resources. As already indicated, some of these early Decrees have been replaced (and these are starred):

Royal Decree of 25 August 1967: relates to the control of foreigners staying and working on drilling platforms.
Royal Decree of 8 September 1972*: set out a framework for offshore operations.
Royal Decree of 3 October 1975*: contained broad general provisions about how the licensee and his contractors were required to carry out their activities. It included requirements for safe systems of work with particular reference to the drilling process.
Royal Decree of 9 July 1976*: made general provisions for safe practices in relation to the production of submarine resources. It applied to the designing, building, installation and operation of production installations, pipe-line systems and shipment installations that were located in a fixed position on or above the sea-bed. This Decree could be applied in areas outside the Norwegian sector of the shelf if such application followed from specific agreement with a foreign state or from international law.

It seems appropriate to refer to the three Decrees which have been repealed, both to indicate the way in which Norwegian law has developed and also because they formed the framework within which more detailed provisions were made.

Until 1985 they were the foundation of the Norwegian offshore safety regime. They were detailed compared with the British legislation and overly complex in the distribution of inspection power and duties between the various Norwegian governmental departments. Possibly one reason for their complexity was that they operated in the context of an 'approval' system, whereby the operator had to present his detailed plans for positive vetting before he received permission to operate in the Norwegian sector. A second explanation is that the Norwegian legislative standards were perceived as administrative standards and, unlike the British Regulations, did not bear direct criminal sanctions – although certain violations of these regulatory standards might be prosecuted under the general criminal code, particularly in cases where risk of personal injury had been created. In certain instances, prosecution under the criminal code was expressly provided for within the Regulations.[30]

An even greater justification for referring to them here is that they are largely reflected in the 1985 legislation and, as has been stated, even though the Decrees themselves no longer have the full force of law, Regulations made under them remain in force for the present. The older laws are also in some instances still relevant in respect of licences created while they were fully in force.

The framework Acts and Royal Decrees of the initial period of Norwegian offshore development were fleshed out by Regulations. The principal safety requirements largely fell into two categories: those which were enforced, up to 1 July 1985, primarily by the Norwegian Petroleum Directorate; and those where the primary responsibility for enforcement lay with the Maritime Directorate. After 1 July 1985 the authority under the petroleum legislation is delegated in total to the Norwegian Petroleum Directorate. Regulations issued by the Maritime Directorate are now enforced on the Norwegian continental shelf by the Petroleum Directorate. The present position is that safety laws for petroleum activities on the Norwegian sector of the continental shelf, with the enforcement of which the Petroleum Directorate is now principally concerned, are set out in their publication: 'Acts, regulations and provisions for petroleum activity' (Volumes I and II).

Volume I contains the new Act of 22 March 1985 on petroleum activities, and the relevant Royal Decrees which are pertinent to petroleum activities. In particular, it gives the three new Royal Decrees of that year which supplement the Act:

Royal Decree of 14 June 1985 supplementing the Act.
Royal Decree of 28 June 1985 making Regulations concerning safety.
Royal Decree of 28 June 1985 concerning the licensee's internal control.

There is also a new framework for enforcement which is contained in:

Royal Decree of 28 June 1985 concerning supervisory activities in the petroleum industry.

The Act and these Royal Decrees largely replace the earlier framework legislation set out above and are significant in respect of the roles envisaged for the Petroleum and the Maritime Directorates in the new system.

The remainder of Volume I gives other framework legislation, such as the Act of 4 February 1977 relating to worker protection and working environment, and the Royal Decree of 25 November 1977 relating to hygiene, although the latter is the responsibility of the Health Directorate. Much of this basic legislation, although emanating from other enforcement agencies, is now applied offshore under agency arrangements, by the Petroleum Directorate. Some of the Regulations contained in Volume I pertain to matters beyond the scope of this book. The relevant provisions for the present purposes are:

Regulations for safety delegates and working environment made by Royal Decree of 29 April 1977.
Regulations relating to worker protection and working environment made by Royal Decree of 1 June 1979, as amended on 13 September 1985.
Regulations for the government Action Control Group in the event of accident leading to or involving risks of extensive oil pollution made by Royal Decree of 19 November 1982.
Regulations concerning labelling, sale, etc of chemical substances made by Royal Decree of 26 November 1982.

Volume II gives the Regulations specifically made for application offshore. These are enforced offshore by the Petroleum Directorate unless otherwise indicated. These are Regulations which are currently being updated, but for the time being they are in the form in which they were authorised by the framework legislation which has now been superseded by the laws of 1985. They are:

Regulations related to safety manning in the event of an industrial dispute of 19 March 1982.
Regulations concerning stand-by vessels of 28 December 1983.
Regulations concerning medical examinations of employees of 1 August 1980 (Directorate of Health).
Regulations covering potable water systems of 23 October 1978 (Directorate of Health).
Provisional Regulations for state-registered nurse on production installations of 11 May 1978 (Directorate of Health).
Standard instructions for physician in charge of fixed installations of May 1982 (Directorate of Health).
Regulations for the structural design of loadbearing structures intended for exploitation of petroleum resources of 29 October 1984.
Regulations for mobile drilling platforms with installations and equipment of 10 September 1973.
Regulations for cranes on production installations of 25 April 1977.
Regulations on deck cranes on drilling units of 31 January 1978.
Regulations for production and auxiliary systems on production installations of 3 April 1978.

Regulations for fixed means of access, etc on production installations of 2 April 1979.

Regulations on arrangements on and below deck on drilling units and for safety measures of 31 January 1978.

Regulations for transfer of personnel to and from production installations, etc of 2 April 1979.

Regulations for drilling for petroleum of 23 September 1981.

Regulations for qualification requirements for drilling personnel of 22 February 1983.

Provisional regulations for diving on the Norwegian continental shelf of 1 July 1978.

Provisional regulations for living quarters on production installations of 2 April 1979.

Regulations concerning the construction and equipment of living quarters on drilling units of 11 June 1982.

Temporary Regulation for electric equipment of installations of 26 July 1985.

Regulations on life-saving appliances on fixed installations of 3 February 1982.

Regulations on approval of survival suits of 10 November 1980.

Regulations for marking of production platforms, pipe-lines and loading installations of 1 December 1976.

Regulations relating to the installation and use of radio equipment of 9 September 1980.

Regulations for helicopter decks of 18 April 1973.

A number of the above Regulations, for example those relating to helicopter decks, were promulgated by enforcement agencies other than the Petroleum Directorate, and expertise lies with that authority (in the case cited, the Aviation Authority), but enforcement is nevertheless primarily through the Petroleum Directorate, who will act with the advice of the specialised agency wherever necessary.

Working Environment Act
Finally, it should be noted that the framework legislation listed in Volume I includes a modified form of the onshore protective employment legislation which was originally extended offshore for fixed platforms by Royal Decree of 1 June 1979, but has been referred to here as amended by Royal Decree of 13 September 1985.

The Maritime Directorate, which formerly had considerable responsibility for enforcement on mobile installations of laws pertaining to petroleum activities, is now only concerned with those aspects of the control of mobile installations which are under Norwegian maritime law. Mobile installations are also subject to Norwegian shipping laws if they are registered as Norwegian ships. Shipping laws cover mobile installations in relation, for example to construction, accommodation standards, and manning and working hours. These Regulations constitute minimum standards for operation of Norwegian-registered installations even when they are operating beyond the Norwegian continental shelf. Compliance with Norwegian standards will not, of course, exempt the installation from other standards required by the coastal state. The Maritime Directorate has published the laws which are applicable to mobile installations in its *Red Book*.

Onshore legislation

The concept that the Norwegian sector of the continental shelf is an extension of the kingdom of Norway facilitates the use offshore of onshore standards to a greater extent than is possible in the British sector. Thus Regulations for electrical installations and explosive substances are more readily invoked in determining standards for use offshore in Norway than is the case in the British sector.

References

1   There were four agreements: 1) The Convention on the Territorial Sea and Contiguous Zone 1965, UKTS 3, Cmnd 2511; 2) The Convention on the High Seas, 1963 UKTS 5, Cmnd 1929; 3) The Convention on Fishing and the Conservation of the Living Resources of the High Seas, 1966 UKTS 39, Cmnd 3208; 4) The Convention on the Continental Shelf 1964 UKTS 39, Cmnd 2422.

2   See *The North Sea Continental Shelf Case* (1969), ICJ Rep 3. When the UK ratified the Convention in May 1964 it was the 22nd ratification. There are now over 50 states bound by it, including most of the major coastal states. Australia and the USA were early signatories.

3   In the case of Britain, the concept of territorial waters first received statutory recognition in the Territorial Waters Jurisdiction Act 1878 which merely gave Britain jurisdiction over foreign ships in British territorial waters. British law does not normally extend into territorial waters. Britain has hitherto exercised its jurisdiction over a three-mile area: this region has been dealt with specifically under the Continental Shelf Act and other offshore legislation, eg The Health and Safety at Work, etc Act 1974 (Application outside Great Britain) Order 1977 Reg 7. Currently it is proposed to extend UK territorial waters to 12 miles.

4   See 2 above.

5   Agreement relating to the exploitation of the Frigg Field Reservoir 1976, Cmnd 6491; agreement relating to the transmission of petroleum by pipe-line from the Ekofisk Field to the United Kingdom 22 May 1973, Cmnd 5432; and Anglo-Norwegian Operators' Agreement relating to Murchison Field: see *Financial Times Survey* 17.9.1979.

6   See Ownership of the Territorial Sea and the Continental Shelf of Australia, Peter Goldsworthy *50 Australian Law Journal* p184.

7   *New South Wales and Ors v Commonwealth of Australia (1975)* 50, ALJR 218.

8   See Ocean Ranger report: now settled in Canada's favour by the Canadian Supreme Court in *Newfoundland Continental Shelf Reference* 23, International Legal Materials 288 (1985).

9   Daintith T and Willoughby GD *UK Oil and Gas Law* Oyez Publishing, 1977 ed, p177.

10 See 1 above.

10a U.N. Document A/CONF. 62/122. of October 7, 1982.

11 Harris *Cases and Material on International Law* 3rd ed, 1983, at p286.

12 S1(1).

13 Now defined in Oil and Gas (Enterprise) Act 1982 S24.

14 See ibid, S21 for present authority for establishment of safety zones.

15 S7 see also S4 (Pt II of the Coast Protection Act 1949); S6 (Wireless Telegraphy Act 1949); S8 (Submarine Telegraph Act 1885 S3).

16 But see Glanville Williams, Venue and Ambit of the Criminal Law, *81 LQR* 1965, p417; and contrast Alex Samuels, The Continental Shelf Act 1964, *BJ of I* 1965, p.155 *et seq.*

17 For further detail see chapter 5.

18 See *Treacy* v *DPP* [1971] AC 537 per Lord Reid at p551 and per Lord Morris at p522; *Cox* v *Army Council* [1963] AC 48 per Lord Reid at p70; *R* v *Martin* [1956] 2 QB 272 per Devlin J at p288; *R* v *Naylor* [1961] 2 All ER 932 per Parker LCJ at p933; and *Reg* v *Jameson* [1896] QB 425 per Lord Russell of Killowen CJ at p430.

19 Contrast S9 of the Australian Petroleum (Submerged Lands) Act 1967 which gives offshore relevance to all state laws which are pertinent to petroleum activities. The Norwegian Petroleum Act of 1985 also intends that all relevant onshore laws shall be applicable offshore.

20 The same reservations must apply to the Police and Criminal Evidence Act's provision that 'installations' are within the interpretation of 'premises' for the purposes of defining police powers of entry.

21 S2 of the Continental Shelf Act.

22 See Employment (Continental Shelf) Act 1978.

23 The development of British regulatory legislation has made no attempt to annex petroleum laws to flag laws. Britain does not, therefore, attempt to regulate the petroleum activities (as opposed to the shipping activities) of its flag ships operating elsewhere than on the British continental shelf.

24 *Report of the Inquiry into the Causes of the Accident to the Drilling Ship Sea Gem*. Cmnd 3409, 1967.

25 For meaning of pipe-line see S33 as amended by S25 of the Oil and Gas (Enterprise) Act 1982.

26 Royal Decree of 14 June 1985 – Regulations supplementing the Act pertaining to petroleum activities; Royal Decree of 28 June 1985 – Regulations concerning safety; Royal Decree of 28 June 1985 – Regulations concerning the licensee's internal control.

27 The regulatory role of the Maritime Directorate is now confined to Norwegian-registered ships.

28 A Norwegian had stolen equipment from an American fixed platform situated on the Norwegian continental shelf. The court of first instance held the shelf was outside the jurisdiction but on appeal the Supreme Court held platforms on the shelf, 'had such a close relationship to Norway ... that when it comes to punishable acts these platforms must be regarded as part of the realm'.

29 The Act was amended by Act No 21 of 25 March 1977.

30 See now Ss55, 57 and 66 of the Petroleum Act of 22 March 1985.

3

The Relationship between Petroleum and Shipping Activities

Principal legislative provisions

The Geneva Convention.
The Jamaica Convention.

British legislation
The Continental Shelf Act 1964.
The Mineral Workings (Offshore Installations) Act 1971.
Petroleum and Submarine Pipe-lines Act 1975.
Merchant Shipping Acts 1894–1984.
The Health and Safety at Work, etc Act 1974.
Oil and Gas (Enterprise) Act 1982.
The Health and Safety at Work, etc Act (Application outside Great Britain) Order 1977.
The Offshore Installations (Registration) Regulations 1972.
The Offshore Installations (Managers) Regulations 1972.
The Offshore Installations (Inspectors and Casualties) Regulations 1973.
The Offshore Installations (Construction and Survey) Regulations 1974.
The Offshore Installations (Emergency Procedures) Regulations 1976.
Diving Operations at Work Regulations 1981.

Norwegian legislation
Maritime Act 1893.
Seaworthiness Act 1903.
Working Hours Act 1977.
Act pertaining to petroleum activities of 22 March 1985.
Regulations concerning safety relating to petroleum activities made by Royal Decree of 28 June 1985.
Regulations relating to worker protection and working environment made by Royal Decree of 1 June 1979.
Regulations concerning exploration and drilling made by Royal Decree of 29 August 1975.
Provisional Regulations concerning diving of 1 July 1978.

Regulations concerning manning of mobile installations of 23 March 1982.
Regulations concerning qualification requirements for personnel on mobile installations of 23 March 1982.
Regulations concerning stand-by vessels of 28 December 1983.
Regulations on life-saving appliances on fixed installations of 8 February 1978.
Regulations for construction and operation of Norwegian mobile installations of 13 January 1986.

The problems of interpretation of the limitations on the coastal state's jurisdiction, which Article 2.1 of the Geneva Convention imposes by granting jurisdiction only for the purposes of exploring for and exploiting natural resources, have already been identified.[1] Most of this work will be concerned with investigating the ways in which coastal states have developed offshore jurisdictions within this framework for the specific purpose of exploiting petroleum resources. However, petroleum laws have been developed to operate in parallel with the traditional systems operated by flag states for the control of their merchant shipping fleets. Thus, a second theme which will run throughout the work will be the interrelationships between shipping and petroleum laws. Problems which will have to be considered from time to time are:

1. What is an installation and how, if at all, does it differ from a ship?
2. To what extent, if at all, does a ship which engages in petroleum activities become subject to petroleum laws?
3. To what extent can the laws of the flag state contribute to safety in petroleum activities?
4. To what extent can the activities of a ship which is not engaged in petroleum activities be subjected to the offshore petroleum laws of a coastal state?

This chapter will consider the conceptual framework, drawing primarily upon British and Norwegian laws to discuss the distinctions between 'ships' and 'installations'. It will also identify the instances where, and the extent to which, petroleum laws have acknowledged and provided for the relationships between petroleum and shipping activities.

The Geneva Convention: petroleum and shipping activities

When the Geneva Convention established a framework for the development by coastal states of marine petroleum resources on the continental shelf, Article 2 employed the following words:

2.1 The coastal state exercises over the continental shelf sovereign rights for the purpose of exploring and exploiting its natural resources.
2.2 The rights conferred in paragraph 1 of this Article are exclusive in the sense that if the coastal state does not explore the continental shelf or exploit its natural resources, no one may undertake these activities, or make a claim to the continental shelf, without the express consent of the coastal state.

The continental shelf was defined in Article 1 so as to exclude territorial waters.

Thus the Convention related to areas which had traditionally formed part of the 'high seas'; it raised immediate questions concerning the accommodation of the activities of exploration and exploitation within the traditional perception of the status of the high seas as regions beyond national jurisdictions, subject to international laws for the protection of nationally-registered ships.

Traditional uses of the high seas

Use of the high seas by persons operating beyond the circumstances in which they could claim the protection of maritime laws has always been regarded as piracy, which by its very existence suggests a defiance of international law. Those who engage in such activities are denied the protection of law; they are likely, by acting in competition with those who enjoy the protection of international law, to find themselves in conflict, possibly even armed conflict, with those whose activities are so protected.

The Geneva Convention elevated exploration for and exploitation of the sea's petroleum resources from the potential status of piracy to something which could be undertaken with national and international protection, providing it was conducted within the terms of the Convention. The parallel with shipping can further be made in that exploration and exploitation would, under the Convention, only be legalised if authorised by a state. However, while the ship's owner is normally free to choose the state in which he registers his ship, thereafter sailing the high seas under the protection of the 'flag' of the state of his choice, he who would exploit the petroleum resources of a continental shelf can only do so if he is authorised by the coastal state in whose sector he intends to operate.

The Convention had, however, not only to legalise the activities of exploration and exploitation as such, but also to provide for the relationship of those activities to other lawful activities, particularly to those which are encapsulated in the concept of freedom of the high seas. It did this in three ways: first (as stated in Article 2.1 above), by granting the coastal state sovereign rights offshore only to the extent necessary for the purposes of exploration and exploitation of hydrocarbon and other mineral resources;[2] second, by protecting other lawful maritime activities from interference by those engaged in the activities of exploration and exploitation; and third, by providing for the operation of 'installations' by coastal states as the focal points of offshore activity. The second and third of these matters are covered by Articles 5.1 and 5.2:

5.1 The exploration of the continental shelf and the exploitation of its natural resources must not result in any unjustifiable interference with navigation, fishing or the conservation of the living resources of the sea, nor result in any interference with fundamental oceanographic or other scientific research carried out with the intention of open publication.

5.2 ... the coastal state is entitled to construct and maintain or operate on the continental shelf installations and other devices necessary for its exploration and the exploitation of its natural resources, and to establish safety zones around such installations and devices and to take in those zones measures necessary for their protection.

The Convention therefore created a new territorial concept in the context of the jurisdiction of sovereign states, namely the 'installation'. The Convention expressly marked the distinction between an installation and land by specifying that installations 'do not possess the status of islands'.[3] Thus there would henceforth be three kinds of legal regime: namely, systems by which sovereign states governed their own territories; systems by which national states controlled offshore petroleum activities in relation to installations; and the internationally recognised rules for the control and use of the high seas.[4]

What is an installation?

The Geneva Convention gives no definition of the word 'installation': it stipulates that an installation shall not have the status of an island, but gives no indication as to its physical, as opposed to legal, attributes. The Convention is silent as to whether an installation is envisaged as mobile, fixed to the sea-bed, floating or submerged. It is merely some 'thing' which is conceived as being necessary for the purposes of exploration and exploitation and is constructed, maintained or operated for those purposes. Moreover, this 'thing' may alternatively be described as 'a device'. The most that can be deduced, therefore, is that an installation is a material object necessary for the activities of exploration and exploitation of natural resources. Since it is a man-made thing it may conveniently be described here as a structure.

It is not entirely clear whether the structure is required to be inherently necessary to the general task being undertaken or whether it is sufficient that it has become necessary because of the particular way in which the particular operation has been planned.

Both British and Norwegian law have avoided defining the word by reference to its physical properties, although this lack of definition appears to be less material to Norwegian law than to British, since it will be recalled that the British Parliament has not indicated in the Oil and Gas (Enterprise) Act 1982 any intention to extend British jurisdiction to the whole of the British sector of the continental shelf but only to installations, certain activities in the vicinity of installations in designated areas, and to pipe-lines.

In both of the national systems many of the special legislative provisions have been expressly devised for the protection and control of installations. For example, the principal regulatory legislation in Britain bears the title the Mineral Workings (Offshore Installations) Act 1971 and many of the Norwegian Regulations have long titles which include the word installation.[5] However, the narrower spatial limits to British jurisdiction, combined with the British system of regulating by legislation which directly incorporates duties which bear criminal sanctions, has led the British to incorporate in their legislation lengthy explanations of what is meant by the word 'installation'. No comparable explanations are to be found in the Norwegian, or – as far as the authors are aware – other systems.

British law

The Continental Shelf Act 1964 made provision for safety zones for the protection of installations[6] and for the application of the criminal and civil law to installations,[7] but failed to provide any guidance as to what was meant by an installation. The Mineral Workings (Offshore Installations) Act 1971 applied to underwater exploitation and exploration of mineral resources, 'from or by means of any floating or other installation' and somewhat circuitously provided:

'Offshore installation' means any installation which is maintained or is intended to be established, for underwater exploitation or exploration to which this Act applies.[8]

The Petroleum and Submarine Pipe-lines Act 1975 S44 extended the application of the 1971 Act to include any installation which:

(a) is maintained or intended to be established, in controlled waters or waters in the United Kingdom, for use in connection with the conveyance of things by means of a pipe constructed in or under the sea; and
(b) is, or when established will be capable, of being manned by one or more persons;

These three major statutes therefore followed the pattern of the Convention and avoided any definition of the physical characteristics of the installation itself, in effect decreeing that anything which was used for the purposes set out in the legislation should be deemed to be an installation. The statutory provisions proved defective, however, because they insufficiently described the activities to which the law was intended to apply. Could, for example, a structure used as dormitory accommodation for those who worked on another structure which was engaged in exploration and exploitation, be deemed to be 'maintained' for underwater exploitation or exploration?

For greater clarification, S24 of the Oil and Gas (Enterprise) Act 1982 provided that for S1 of the Mineral Workings (Offshore Installations) Act 1971 there should be substituted:

(1) This Act shall apply to any activity mentioned in sub-section (2) below which is carried on from, by means of, or on, an installation which is maintained in the water, or on the foreshore or other land intermittently covered with water, and is not connected with dry land by a permanent structure providing access at all times and for all purposes.
(2) The activities referred to in sub-section (1) above are:
   (a) the exploitation or exploration of mineral resources in or under the shore or bed of controlled waters;
   (b) the storage of gas in or under the shore or bed of controlled waters or the recovery of gas so stored;
   (c) the conveyance of things by means of a pipe, or system of pipes, constructed or placed on, in or under the shore or bed of controlled waters; and
   (d) the provision of accommodation for persons who work on or from an installation which is or has been maintained, or is intended to be established, for the carrying on of an activity falling within paragraph (a) (b) or (c) above or this paragraph.

Sub-section 4 provides that 'controlled waters' have a meaning which includes tidal waters up to the limits of territorial waters and waters in any area designated under the Continental Shelf Act 1964.

Sub-section 5 further provides that 'installation' includes:

(a) any floating structure or device maintained on a station by whatever means;

This provision, while expressly stating that an installation need not be fixed to the sea-bed, implies that an installation may be mobile, being capable, that is, of being moved from place to place either under its own power or by virtue of being towed or otherwise propelled by an external use of power.

Application clauses in Regulations made under the Mineral Workings (Offshore Installations) Act 1971 are not standardised. However, the Offshore Installations (Registration) Regulations 1972, and the Offshore Installations (Construction and Survey) Regulations 1972 acknowledge this distinction between 'fixed' and 'mobile installations':

2. (1) For the purposes of these regulations:

'mobile installation' means an offshore installation which can be moved from place to place without major dismantling or modification, whether or not it has its own motive power; and

'fixed installation' means an offshore installation which is not a mobile installation.[9]

Incidentally, it may be noted that if a structure has been registered as an installation, it is likely for all practical purposes to determine that it is actually an installation; however, acceptance of the *de facto* position does not satisfy the underlying legal question whether it ought to have been so registered. It may be noted that there are no authoritative cases on the meaning of the word installation.[10] The lack of authority may be indicative that there is in practice no real problem in identifying whether or not a particular structure is an installation.

What is a ship?

In the British regulatory system, a British-registered ship is subject to the Merchant Shipping Acts 1894–1984 and for most purposes ships – even British-registered ships – are not obviously subject to the regulatory regime which governs petroleum activities in the British sector of the North Sea continental shelf. The same has to be so in other coastal state regimes because of the protection which the Geneva Convention gave to ships to enjoy the freedom of the high seas.[11]

Nor would it appear that the merchant shipping laws of a coastal state have any application to ships other than those registered under its own flag. It is interesting to note that the Merchant Shipping Act 1979 S27 (1) (a) gives wide powers in respect of ships to an inspector appointed under the merchant shipping legislation. It provides:

An inspector appointed in pursuance of Section 728 of the Merchant Shipping Act 1894 –

(a) may at any reasonable time (or, in a situation which in his opinion is or may be dangerous, at any time) –

...

(ii) board any ship which is registered in the United Kingdom wherever it may be and any other ship which is present in the United Kingdom or the territorial waters of the United Kingdom,

if he has reason to believe that it is necessary for him to enter the premises or board the ship for the purpose of performing his functions as such an inspector.

It may be, however, that this provision gives relatively little power to the inspector in relation to foreign-registered ships servicing installations on the British sector of the continental shelf. First, such ships would be likely to be operating outside territorial waters and Britain unlike Norway does not regard her continental shelf as 'within the kingdom'; second, the functions of a British inspector must inevitably be limited in relation to ships flying a foreign flag.[12]

To be governed by the British Merchant Shipping Acts a structure must fall within the definition of a 'vessel' contained within these Acts.[13] The criteria under this legislation is that the structure must be capable of moving itself without aid. Thus, a structure which can only be moved by towing or by carriage on another structure is unlikely to be deemed a vessel within the legislation. There has been a considerable amount of litigation in which it has been in issue whether a particular structure on the high seas ought to be deemed a vessel. Most of the relevant cases have been concerned with determining issues of liability and salvage matters in the wake of a collision.[14] Under S41 of the Merchant Shipping Act 1979, the Secretary of State may, by order, apply the Acts to 'any thing designed or adapted to be used at sea'; this has obvious applications to petroleum activity.

Comparison of the statutory concept of an installation with the statutory concept of a ship shows that one is a structure employed for the purpose of the petroleum activities specified in the Oil and Gas (Enterprise) Act 1982 and the other is a structure which must, according to the Merchant Shipping Act 1894, be capable of being navigated on the high seas. The two regulatory regimes do not recognise that there may be relationships and overlaps between the activities with which they are concerned.

Can a structure be both an installation and a ship?

The classification of structures as ship and installation are not self-evidently mutually exclusive categories: a ship is a structure capable of movement, an installation is a structure engaged in petroleum activities. It is therefore possible that a mobile installation may be a ship or that a ship engaged in the activities specified in the 1982 Act may be an installation.

In British law, while a structure may cease to be an installation when it ceases to be engaged in any of the specified activities, a structure does not cease to be a ship if it is stationary but still capable of propelling itself. An anchored mobile installation carrying out drilling will therefore be both an installation and a ship (if self-propelled). On the other hand, a ship which is anchored for an indefinite period while servicing an installation, for example as a stand-by, will remain a ship. It is questionable whether a floating production installation with its own power is still a ship.

In fact, there are in British law situations where a structure might logically be described as either a ship or an installation. The two situations in which this choice is most likely to occur are: first, where a ship is engaged in drilling; and second, where it is tied alongside an installation and providing accommodation for persons working on the installation. The question which then has to be addressed is whether the structure can, if it has elected to be classified as a ship, resist the regulatory regime governing installations or vice versa. It would appear not: the classification of the structure is determined by its function, not by a management decision as to which regulatory regime it should accept. In spite of the fact that the British regulatory agencies appear to ignore that a structure may have a dual status, there would seem to be no doubt that in law the dual status does at times exist.

There is, however, an important difference between petroleum and shipping law: a structure which is intended for work in petroleum activities on the continental shelf must meet any requirements of the coastal state in whose sector of the shelf it intends to operate as to being of a standard fit for such operation in the waters in question. Moreover, under the 1982 Act if a mobile drilling rig is moving from one drilling location to another on the British shelf it will be an 'installation' within S24(2)(d), since it is within the meaning of either 'has been maintained' or 'intended to be established'; during the time that it is moving it will be providing accommodation for some personnel. However, under S28(1) an installation in transit is to be treated as a vessel. Therefore such a mobile installation may be treated as both a ship and an installation.[15]

There is no requirement that a ship sailing on the continental shelf must register in the coastal state and comply with its merchant shipping laws; nor indeed is it clear that it need register as a ship with any state.

However, there are strong practical reasons for the registration of a ship with a classification society such as Lloyds, for insurance and similar purposes, even if it is already classified as an installation. The wide choice of jurisdiction, and the knowledge that some shipping regimes are more exacting than others, may cause the ship owner to choose to register in one state rather than another. The fact that the adoption of a flag state is bound to impose some constraints on the owner and operator of the vessel is most unlikely to discourage registration. The benefits of registration outweigh the constraints which it imposes: the owner's power to select the flag of his choice enables him to maximise the benefits and minimise the burdens of registration.

Being registered as a ship gives the structure a status, since entitlement to registration depends on capability world-wide rather than present function. On the other hand, entitlement to carry out petroleum activities is an entitlement which is subject to geographical and time boundaries. For example, a mobile installation travelling through 'no man's land' between two continental shelf areas, might find itself devoid of status if it were not registered as a ship, since if it were in transition it might well be divested of the status of installation. Thus, drilling mobiles, which may well travel from New Zealand to Norway in search of work will be likely to be registered as ships and retain that registration throughout their working lives, regardless of whether they are at a given point of time also subject to petroleum laws.

The converse is not necessarily the case. A ship may wish to avoid the regulatory standards which apply to installations, since these standards may differ substantially from those which apply to it when classified as a ship. In the case of mobile installations, the United States requires those registered under its flag to comply with the relevant International Maritime Organisation[15a] (IMO) Conventions, including SOLAS and Load Line. There may well, therefore, be instances in which ships which are servicing the offshore industry resist the application of petroleum laws, on the argument that what they are doing is not within the ambit of the relevant petroleum laws. It was because the status of accommodation barges and mobiles, commonly called 'flotels', was insufficiently clear that they were expressly brought within the ambit of British petroleum laws by S24(2)(d) of the Oil and Gas (Enterprise) Act 1982.

What does dual status imply?

Acceptance that a structure has a dual status implies that the structure has to comply with two regulatory systems: the petroleum regime of the place where it is sited for the time being, and a merchant shipping regime which will remain with it in whatsoever part of the world it happens to be, until such time as it ceases to be entitled to fly the particular flag of the state where it is registered. It is also possible that the flag state might stipulate as part of its petroleum regime that if its ships choose to engage in petroleum activities they must, in relation to these activities, comply with the petroleum standards imposed on installations operating on its own continental shelf. This regulatory technique would ensure minimum operational standards for the vessels flying that flag and be valuable for structures operating in regions where there are no, or insufficient, standards imposed on petroleum activities. It could, however, be embarrassing for the operator if there were irreconcilable conflicts between flag and petroleum standards. This could, in the last instance prevent the operation of structures in certain waters while flying certain flags.

The fact that Merchant Shipping Acts and petroleum laws have developed and been enforced separately by different government agencies in the British sector, means that there is no clear policy of co-ordination between the two regimes applying to British ships which are operating as installations. Thus, the structure must initially comply with the structural standards of both agencies, thereafter maintain those standards and additionally operate to the systems within both sets of legislation.

In practice, a ship which is engaged in petroleum activities will generally have two workforces on board: the ship's crew, under its captain, responsible for the navigation of the vessel, and a second group of workers engaged in petroleum activities. This situation is likely to exist in a mobile installation. In the British system, the ship's crew are governed primarily by the Merchant Shipping Acts and the petroleum employees are governed by the Mineral Workings (Offshore Installations) Act and the Health and Safety at Work, etc Act, which has been extended offshore by the Health and Safety at Work, etc Act (Application outside Great Britain) Order 1977. However, the mineral workings legislation and the Health and Safety at Work, etc Act can also apply to the ship's crew. For example, they apply to the crew of a supply vessel when it is in a safety zone; they could also apply to a mobile installation which had remained subject to petroleum laws while in transit on the British continental shelf.

Command systems

Undoubtedly the fact that the inquiry which followed the capsizing of the mobile installation Sea Gem,[16] recognised the similarity between mobile installations and ships, influenced the form of the remedial legislation – the Mineral Workings (Offshore Installations) Act 1971. The regulatory system built under that legislation clearly treats an installation as a structure similar to a ship and needing a similar control system, both to ensure the integrity of the structure[17] and to ensure an appropriate command structure.[18]

By whatever regulatory system or systems a mobile installation is governed, there is a need to operate to an organisational plan which determines the relationships between the senior managers of the petroleum workers and the senior officers of the ship's crew. In the drilling operation in particular, it is necessary to have a rule as to whether the needs of the petroleum activity or the needs of the ship will have priority in emergency. The British system gives overall command to the installation manager[19] even though there may be on board a person who has responsibilities for maritime matters.

The Canadian commission appointed to conduct an inquiry into the reasons for the sinking of the drilling rig Ocean Ranger off the coast of Newfoundland, not only stressed the importance of having a clear line of authority but considered highest authority should be placed in the person best able to evaluate the needs of the ship against the needs of the drilling exercise. At the time, the management structure had reflected the North American practice of granting precedence to the industrial rather than the maritime crew. The operations which secured stability and safety of the rig were relegated to a subordinate role, comparable with that of any other support group. The precedence of the industrial activity was reflected in the command hierarchy, the inadequate provision made for marine qualified manpower, and in the training policies followed on board. The operators were in the habit of designating the toolpusher as the person in charge when the rig was on location, and the ship's captain as in command when the ship was moving. The toolpusher had no marine qualifications and no experience of ballast control. The fatal weakness of the system proved to be that it fell to the toolpusher to determine whether or not the weather was too severe to permit the drilling operation to continue without endangering the vessel.

Similar difficulties concerning the conflict between maritime and petroleum regimes appear to have been encountered in Australia. The operation of fixed installations is under the state-enforced Petroleum (Submerged Lands) Act, but mobile installations like ships are subject to Commonwealth law – the Navigation Act 1912 – which is policed by the Department of Transport. Until recently, neither regulatory code applied to an installation under tow, as the Key

Biscayne catastrophe demonstrated. Moreover, while the Department of Transport Marine Order (Part 47) established detailed requirements relating to the design and sea-state operation of mobile drilling units, it specifically excluded drilling operations.

In the British system, while overall authority rests with the installation manager he must be competent to oversee both drilling and emergency requirements.

Norwegian legislation
The development of the Norwegian system has avoided some of the problems suggested by the above account of the British system. The Norwegian system did not give any one agency exclusive responsibility for policing installations in their sector of the North Sea continental shelf. Until the Act of 22 March 1985 pertaining to petroleum activities, primary responsibility for mobile installations indisputably rested with the Maritime Directorate, the agency responsible for ensuring the integrity of Norwegian-registered ships. However, it also had a co-ordinating role in approving mobile installations for use on the Norwegian continental shelf, pursuant to both maritime and petroleum legislation. After the passing of that Act, the standards for Norwegian mobile installations are still those of the Maritime Directorate, evidence of whose surveys will be relevant documentation to be considered by the Petroleum Directorate in enforcing standards on the Norwegian continental shelf.

Giving the Maritime Directorate principal responsibility for mobile installations enabled full recognition to be given to the physical and operational similarities between ships and installations, but the involvement of other regulatory systems enabled an open declaration of the total regulatory control which would apply to a ship which was operating as an installation. The publication by the Maritime Directorate of the *Red Book*[20] contained all the Regulations which applied to mobile installations under the Norwegian system, regardless of the enforcement agency which created and enforced the Regulations, but nevertheless gave a clear indication of the importance of merchant shipping laws to installations.

However, the initial Norwegian solution had its own difficulties. Historically, at least, the allocation of responsibilities for mobile installations to the Maritime Directorate, with the Petroleum Directorate principally responsible for day-to-day enforcement on fixed platforms, gave the impression of a less coherent enforcement pattern for petroleum activities than would have been the case had the Maritime Directorate played a more subordinate role. It has yet to be seen whether the increased role given, since the Act of 22 March 1985, to the Petroleum Directorate for the formulation and execution of policies relating to petroleum activities, will in future end the hitherto somewhat sharp distinctions between fixed and mobile installations when operating on the Norwegian continental shelf.

The Norwegian Maritime Directorate's *Red Book* always acknowledged that there are both shipping and petroleum regimes applicable in the Norwegian regulatory system to ships operating as mobile installations. In the revised 1986 version it contains only the Regulations made under the Seaworthiness Act of

1903 applicable to Norwegian mobile installations. A clear account of the situation intended by the new legislation can be found in the proposition to the Storting at the time that the new legislation was formulated.

The outcome, therefore, is that the Norwegian system is now much more coherent, because the Norwegian shipping laws are relevant only to mobile installations which are registered as Norwegian ships. Thus the revised *Red Book* serves as a useful manual indicating the regime to which a Norwegian flag ship must conform when seeking to be approved as an installation operating in the Norwegian sector of the North Sea and elsewhere. It does not totally overcome the problem of the dual system of control if the circumstances are that the ship which is seeking approval as a mobile installation to engage in petroleum activities in the Norwegian sector is a Norwegian-registered ship.

The petroleum regime may, in effect, regulate the working hours and other working conditions of the personnel on a mobile installation if that installation is flying a foreign flag. Section 6 of the otherwise repealed drilling Regulations of 29 August 1975 provides that the whole crew is subject to working hours Regulations[21] if the installation performs exploratory drilling. Section 5 of the safety Regulations of 28 June 1985 resolves the conflict between the two regimes in the case of a Norwegian flag vessel. Section 5 of the Regulations of 28 June, because it requires documentation from the licensee, has the effect of placing the responsibility upon him for ensuring that an installation complies with the necessary standards.

The problem of determining command structures between ship's crew and drilling crew is covered in the following terms in the Norwegian Regulations concerning the manning of Norwegian flag mobile installations, which stipulate that the platform manager:

> has the highest authority on board and is responsible for the stability and safety of the drilling unit/installation during drilling, moves and anchoring.

> shall be familiar with the structural conditions of the drilling unit/installation, and is responsible for ensuring that the structural loads during anchoring or towing operations do not exceed the limit for permissible load on the platform.

> shall, in situations where the safety of the crew of the drilling unit/installation is endangered due to drilling problems or for other reasons, consult with the drilling section leader and the stability section leader before making his decision ...

The Norwegian practice is to require the platform manager of any installation on the Norwegian sector of the continental shelf to be a master mariner with training in stability and ballast control. This requirement is extended to any Norwegian flag installation in whatsoever part of the world it is operating. Moreover, the safety Regulations of 28 June 1985 require that organisational plans and specified communication lines clarify the respective responsibilities of the platform manager and the operator.

Relationship between installations and ships

There are situations in which an installation operates in conjunction with or is serviced by a ship. In these situations it is vital to the safety of both the structure and the personnel, engaged respectively on the ship and on the installation, that provision is made for the integration of the two functions into a joint enterprise. There are three particular situations which have been recognised and provided for within the legislative systems under consideration. They will be noted here and dealt with more fully in Part II of the book. They are:

Supply ships

Ships play a vital role both in the construction and operational phases of installations, in the provision of a link between the installation and land. They are a means of providing the ship with supplies and taking goods from the installation back to land base.

It is beyond question that, when such a supply ship is on the high seas, it is neither an installation nor within the ambit of petroleum laws: it is at the point of loading and unloading that it becomes vitally important that the exercise is a joint one, under a single command, and it is at this stage that petroleum laws have sought to impose controls over ships. That the concept of a single chain of command at this stage shall be a legal requirement is provided for in both the British[22] and the Norwegian legislation; in both systems command rests with the installation manager, as far as the safety of the installation is concerned, but similar authority rests with a ship's captain for the safety of his vessel, so there is a theoretical potential for conflict of authority.

Stand-by vessels

Both British and Norwegian law require that every manned installation shall be attended by a stand-by vessel capable of rendering assistance in an emergency.

In neither system do these provisions of the petroleum laws bring stand-by vessels completely within their ambit. A licensee's decision to employ a stand-by which did not meet the requirements of the petroleum inspectorate would place the licensee in breach of his duties.[23] In effect, he would be deemed to be operating without a stand-by; he would then have the option of requiring the ship he had chosen to be brought up to standard or seeking another vessel. In this matter he would presumably be influenced both by his contractual arrangements with the ship in question and the extent to which it fell short of the requirements of the petroleum laws. In no instance, however, could the ship's captain or owner be proceeded against directly under the British mineral workings legislation. In the Norwegian sector, on the other hand, the main rule, which is contained in the safety Regulations of 28 June 1985, S5.2, is that all ships, vessels, or barges performing petroleum activities in relation to installations (fixed or floating) are covered by the Petroleum Act as to these functions.

Diving Operations

Diving operations in the vicinity of an installation may well be conducted from a ship alongside the installation rather than from the installation itself. This then creates another situation in which there is a need to co-ordinate the activities of the installation and the ship, both for the safety of the ship and the installation as structures and for the safety of the personnel involved. This matter is again dealt with explicitly in both the British and the Norwegian systems. Moreover, in the Norwegian sector, the general provision noted above – that ships engaged in petroleum activities are brought within petroleum laws for the purposes of these activities – applies here as well as to stand-bys, etc.

Flag and shelf standards

In certain parts of the world, offshore petroleum activity may be carried on in locations where petroleum or shelf safety standards are inadequate, unsuitable or non-existent. In such a situation any applicable flag standards may assume considerable importance.[24] Flag legislation regulating the seaworthiness of shipping is now commonplace. As well as providing for structure and equipment standards, such legislation frequently covers manning, qualifications and training of crews.[25] In some instances – in Britain, for example – the legislation has been extended to include onshore-style safety standards, such as safe machinery and systems of working.[26]

Certain countries which consciously seek to regulate mobile drilling units under their flag as ships, have taken these developments to their logical conclusion by regulating, under shipping law, for the special petroleum activities of drilling units, including industrial as well as maritime risks; but with the significant distinction from specific petroleum activity legislation that the shipping standards apply to mobile drilling units on a world-wide basis.

The United States Department of Transportation requirements for American-registered mobile drilling units, issued in January 1979,[27] provide an interesting example of this development. Apart from ship-like construction and stability standards, the requirements lay down minima for life-saving, fire-fighting, cranes, mobile cranes and trucks, accommodation and welfare; also requirements for industrial systems and for periodic inspection and testing of machinery and apparatus. There are no express requirements for the manning of the unit in the drilling mode; in the navigation mode the manning standards relate to those applicable to US-registered ships generally.

Canada regulates all vessels registered under its flag by the standards of the Canadian Shipping Act, which also applies to all vessels of whatever registration operating within 12 miles of its shores. Canadian shipping standards are confined to ship-like criteria, and not to the special characteristics of mobile drilling units. On the Canadian continental shelf, primary legislation regulating the activities of mobile drilling units is contained in the Canada Oil and Gas Drilling Regulations 1980 which, as petroleum legislation, is applicable to units of any nationality. It would be true to say that, in enforcing the latter Regulations, the Canada Oil and Gas Administration did not regard the marine safety of units as a priority, but concentrated upon the drilling programme. In fact, in Canada as elsewhere, the 'manning' requirement that most appears to concern the regulating authorities is the nationality of the crews of foreign units allowed to work upon its continental shelf.[28]

Norway, on the other hand, in its system (already discussed), appears to have taken the best features of the flag system, and, where appropriate, combined these with aspects of petroleum laws in regulating mobile drilling units and other ship-like structures, although there are still problems which have to be resolved. While making special provision for mobile units as ship-like structures under the powers of its basic shipping legislation, Norway was not unmindful of the problems created by foreign flag mobiles operating on its own continental shelf. Clearly the latter could be regulated within the ambit of the Norwegian petroleum or shelf legislation: nevertheless Norway has seen fit, on occasion, to blur the flag and shelf powers when legislating for mobiles.

On the basis of this analysis, the British system appears relatively simple. Flag standards relating to structure, manning, and operational safety apply to British shipping operating world-wide, including shipping engaged in petroleum operations. Up to the present, it appears that no flag standards exclusively relating to ships engaged in petroleum activities have been made. 'Petroleum' standards relating to all installations, including mobiles, have been made, with application to British or foreign mobiles operating on the British continental shelf.[29]

An evaluation of the potential of flag legislation for achieving safety in the petroleum industry offshore might be thus: flag standards are most useful in situations where shelf standards are lacking or inadequate. In such situations, ship or installation operators might minimise their safety commitment by registering their 'vessel' with countries that have not involved themselves, through their legislation, in offshore safety matters. Even where shelf standards are comprehensive, as in Britain or Norway, the enforcing authority might be reluctant to enforce all their shelf standards on the internal operations of a foreign flag installation or vessel. This renders the existing flag standards applicable to that particular unit, for example on working hours, of crucial importance.

Inasmuch as flag standards apply world-wide, their potential for assisting in the raising of labour and safety standards is very great, serving as they do to introduce the standards of the developed world to offshore activities carried on in the less developed parts of the world. This consideration is reflected in the work of the International Maritime Organisation (IMO), which was set up (under a different title) by the United Nations in 1958. Its purpose was to develop minimum standards for all vessels operating in international waters. With this objective it seeks to build upon the best of existing flag standards.

Members of the IMO, who formulate these standards, bind themselves to incorporate them within their own flag legislation. Certain of the IMO standards for general shipping, such as the Safety of Life at Sea Convention 1960, and the Load Line Convention 1966, have been applied to mobile installations through national shipping legislation.[30] Thus the Ocean Ranger[31] was approved by the US coast guard under US flag legislation as being in compliance with IMO standards, as incorporated in US shipping legislation. It is significant that, since the report on the sinking of the Ocean Ranger was published, both Canada and Norway have proposed to the IMO that uniform standards for certificates of competency for crews of mobile drilling units should be promulgated, particularly relating to the training and qualifications of installation managers and other key marine personnel on board, such as stability officers.[32] The British Department of Energy has published a Guidance Note on the latter subject[33] and this standard is being adopted by the IMO.

Other shipping activities

The Geneva Convention specifically sought to preserve the freedom of the high seas in so far as possible. There can be no doubt, therefore, that the coastal states have no general power to control the movement of, or conditions of employment within ships, supply ships included, which operate within their sector of the continental shelf. The Convention did, however, as is noted above, give coastal states power to create safety zones round installations and, 'to take in those zones measures necessary for their protection'.[34]

Both the British and the Norwegian regulatory systems have taken advantage of the concept of safety zones. The British Continental Shelf Act S2 initially enabled the then Ministry of Power (now the Department of Energy) to create safety zones by statutory instrument for individual installations. This provision has now been replaced by S21 of the Oil and Gas (Enterprise) Act which enables the Secretary of State to establish zones more speedily by a less formal procedure and also spells out more fully the implications of creating a safety zone. The new provision is:

> A vessel shall not enter or remain in a safety zone except under and in accordance with the terms of an order made or consent given by the Secretary of State.[35]

It is further provided that if a vessel enters or remains in a safety zone contrary to this provision, the owner and the master may both be liable to prosecution for a criminal offence.[36]

These provisions will shortly be replaced by sections 21–24 of the Petroleum Act 1987, which will provide a simplified procedure for the establishment of safety zones, and for offences connected therewith.

Section 47 of the principal Norwegian Act of 22 March 1985 declares that there will be a safety zone surrounding every installation, and enables safety zones to be created a reasonable time before the placement of the installation. The procedure for creating safety zones appears to be simpler under the Norwegian system; this and the fact that Norway exercises jurisdiction over the whole of her sector of the shelf, whereas Britain purports to exercise jurisdiction only in designated areas, combine to make it simpler to ensure that every installation has a safety zone in the Norwegian system.

In neither system do installations have safety zones when in transit, since at that stage they are not involved in petroleum activities. However, the Norwegian legislation enables a safety zone to be established prior to the placement of the installation, in order to secure the position for the installation.[37] Where an installation is sited at the boundary of the coastal state's sector its safety zone may cross the boundary. The Norwegian view (likewise the British) is that in these circumstances they are entitled to extend their laws to the limits of the zone, albeit this may be in another coastal state's jurisdiction.[38]

In both jurisdictions it is assumed that the safety zone will be enforced by the installation manager who will admit to the zone, when it is safe for them to be admitted, supply ships and the like which have business with the installation. Unauthorised entry into a zone could, in the last instance, be an act of aggression which might be repressed by force. Most violations of safety zones are, however, likely to be caused by negligent navigation rather than wilful disregard of the installation's legal entitlement. In such instances effective sanctions may be hard to impose.

The penal provisions of the British law would seem to be unenforceable unless the offending vessel brought itself within the jurisdiction of the British courts and a foreign flag ship would be unlikely to use a British harbour after such an incident. Nevertheless there have been two successful prosecutions taken in British courts for violation of safety zones; in one instance the vessel was foreign-registered but its owners answered the summons.[39]

Emergency procedures

In many ways the problems of evacuation in an emergency will be the same on an installation as on a ship. In both instances there is the same need to have contingency arrangements for abandonment of the structure and to ensure that there is adequate life-saving equipment available for use. Not surprisingly, therefore, the petroleum regimes have adopted the systems of drills and musters which are generally used on board ship,[40] use the same types of survival equipment, and call on maritime authorities to inspect and approve such equipment when it is used on installations. Whether or not this approach is always relevant to installations is open to question.

Summary

It would seem that the following assertions can be made about the relationships between shipping and petroleum activities on the continental shelf:

1. A structure can simultaneously be both a ship and an installation and have to comply with both shipping and petroleum laws.
2. A coastal state cannot impose its merchant shipping laws on other than its own registered ships.
3. A state can impose its petroleum laws on mobile installations which are also its flag ships, in whatsoever part of the world they are operating, but an installation may not escape the local petroleum laws by electing to comply with conflicting requirements imposed by the flag state. Thus its flag requirements might disqualify it for work in another regime.
4. A coastal state cannot regulate for the control of the ship's crew on foreign installations operating in its sector. It could, however, refuse permission to operate to an installation with whose crew conditions it was not satisfied. More generally, a coastal state may, through its licensing procedure, directly or indirectly, exercise considerable control over installations and shipping of any nationality utilised by its licensees in petroleum activities.

5. At the interface between activities by supply ships, diving ships and stand-bys on the one hand and installations on the other, it is desirable to have systems of work which give ultimate control to either the captain or the installation manager: a dual control system is liable to lead to conflict and confusion. The same considerations apply as between drilling operations and navigation.

6. Similarities between installations and ships have caused emergency procedures for evacuation of installations to follow the patterns adopted for ships.

References

1 S2.
2 This book is concerned only with hydrocarbons, namely oil and gas, described as petroleum.
3 Article 5.4. Presumably the reason for this was that otherwise the existence of an installation under the control of a sovereign state might extend that state's rights as a coastal state; also the installation might claim territorial waters.
4 For the likely effect of the Jamaica Convention, see Chapter 2.
5 Eg Provisional Regulations for living quarters on production installations of 2 April 1979.
6 Chapter 2.
7 S4
8 S1
9 Reg 2(2) in both sets of Regulations provides: 'Nothing in these regulations shall apply to dredging installations which are registered as vessels.'
10 See *Elms* v *Foster Wheeler Ltd* [1954] 2 All ER 714; *Engineering ITB* v *Foster Wheeler Ltd* [1970] 2 All ER 616; *Prestcold (Central) Ltd* v *Ministry of Labour* [1969] 1 All ER 69; and *Procurator Fiscal* v *Union Oil Co of GB* [1978] *Unreported* for such guidance as is available
11 The exceptions are that all ships must observe the safety zone which surrounds an installation and if a ship is engaged as a stand-by for an installation it will in effect have to comply with the special Regulations pertaining to stand-bys. For these matters see below.
12 But Regulations made under the Mineral Workings (Offshore Installations) Act, ie The Offshore Installations (Inspectors and Casualties) Regulations 1973, may enable inspectors to board and inspect the vessel if it has become an installation.
13 Merchant Shipping Act 1894 S596.
14 See Summerskill M *Oil Rigs – Law and Insurance* Stevens, 1979.
15 Oil and Gas (Enterprise) Act 1982 S28(1); *contra* Norwegian Petroleum Act 1985 S1. But in British law there is no requirement for an installation manager. See the Offshore Installations (Manager) Regulations Reg 5. Also for the purposes of certification of fitness (see Chapter 6) Britain has always attempted to control mobiles on entering controlled waters.
15a Discussed, post, p. 63.
16 *Report of the Inquiry into the Causes of the Accident to the Drilling Rig Sea Gem* HMSO, Cmnd 3409, 1969.
17 The Offshore Installations (Construction and Survey) Regulations 1974.
18 In particular, compare the status of a ship's captain with that of the installation manager as provided in the Mineral Workings (Offshore Installations) Act 1971 S5 and the Offshore Installations (Managers) Regulations 1972.
19 The Mineral Workings (Offshore Installations) Act 1971 S5(4).
20 Flyttbare boreplattformer. Now Regler for Flyttbare Innretninger 1986.
21 See Regulations relating to worker protection and working environment S2. Under S2.2 of the Royal Decree of 1 June 1979, it is deemed possible to regulate ships' crew conditions of even foreign vessels, in accordance with these regulations if they are so low as to be totally unacceptable, or if Regulation is necessary for safe performance of petroleum activities.
22 S5 of the Mineral Workings (Offshore Installations) Act 1971.

23 The British Regulations in fact place the duty upon the installation owner and the concession owner (Reg 11).

24 Britain: Merchant Shipping Act 1894, as amended; Norway: Act for the Public Control of Seaworthiness of Ships dated 9.6.1903, as amended.

25 Britain, see Merchant Shipping (Certification and Watchkeeping) Regulations 1982 SI No 1699.

26 Merchant Shipping Act 1979 S21.

27 46 CFR, Parts 50–61 and 107–113.

28 *Report of the Royal Commission on the Ocean Ranger Disaster* Canada, August 1984, pp9, 35, 37.

29 See Chapter 2.

30 See US MODU Regulations – 46 CFR 107.401; Australian shipping laws have also followed IMO standards.

31 See Ocean Ranger Report.

32 IMO, Marine Safety Committee; 51st Session, Agenda Item 20, 27 March 1985.

33 Training Guidance Note No 4 *Guidance on the Selection and Training of Ballast Control Operators* These standards are being adopted by the IMO.

34 Article 5.

35 S21(3).

36 S21(4).

37 S47 of the Act of 22 March 1985.

38 Odelsting Proposition No 72 at p 143. (Ministry of Petroleum and Energy, 1982–83).

39 See also *Champion Tankers Ltd* v *Norpipe A/S and Others* [1984] AC 563. No point of jurisdiction was taken where the owners of a Liberian-registered tanker appeared before a British civil court and were found liable for fouling a Norwegian pipe-line off Hartlepool.

40 British: the Offshore Installations (Emergency Procedures) Regulations 1976; Norwegian Regulations on life-saving appliances on fixed installations of 8 February 1978 and Regulations for mobile installations of 10 September 1973.

4

Responsibilities for Ensuring Safety Offshore

Principal legislative provisions

British legislation

Merchant Shipping Act 1970.
Mineral Workings (Offshore Installations) Act 1971.
Health and Safety at Work, etc Act 1974.
Petroleum and Submarine Pipe-lines Act 1975.
The Offshore Installations (Operational Safety, Health and Welfare) Regulations 1976.
Lead at Work Regulations 1980.
The Diving Operations at Work Regulations 1981.
Petroleum (Production) Regulations 1982.

Norwegian legislation

Act relating to the public control of the seaworthiness of ships 1903.
Worker Protection and Working Environment Act 1977.
Act pertaining to petroleum activity 1985.
Regulations concerning safety in petroleum activities of 28 June 1985.
Regulations concerning the licensee's internal control made by Royal Decree of 28 June 1985.
Regulations for mobile drilling units of 10 September 1973 and other Regulations and Guidelines in Appendix 1 and 2 of Chapter 6†
Norwegian Penal Code

Many jurisdictions with an interest in petroleum and shipping laws have taken the fundamental decision to enforce their chosen legislative standards in these

---

† A difficulty here, and throughout much of the remainder of the book is that in 1986 the Maritime Directorate reissued Regulations relating to Norwegian mobile installations, in a revised edition of the *Red Book*, giving them authority under the Seaworthiness Act of 1903. Thus, while the standards recognised by the two enforcement agencies are largely the same, the reference is different: the Petroleum Directorate's appendices use the former references. Eg the quoted Regulations for mobile drilling units of 10 September 1973 are now the Maritime Directorate's Regulations of 13 January 1986.

fields by the use of criminal sanctions. This decision has had inevitable consequences when it is sought to enforce these legislative standards against the complex organisations that have, of necessity, evolved for the exploitation of offshore petroleum resources.

Although in most jurisdictions criminal liability has developed as a means of controlling individual anti-social conduct, its use as a regulatory sanction requires the criminal law to be applied in the following typical situations:

1. Against individuals for their personal and intended contravention of regulatory standards.
2. Against individuals for an objectively assessed failure to comply with legal obligations imposed upon them, either as members of a group (eg as employees), or imposed upon persons generally.
3. Against individuals whom the law holds responsible while acting in some special capacity. Liability may be imposed on them for their personal fault or for the fault of others: either for the identified failings of subordinates to whom they have delegated the discharge of their statutory obligations, or for the failure of a 'team' to accomplish statutory objectives (eg the installation manager, or captain of a vessel are persons upon whom such responsibility may be imposed).
4. Against individuals, or more probably corporate bodies, for a failure to comply with obligations imposed upon them in some capacity such as 'employer', 'licence holder', or 'owner of a vessel or installation'.

Britain has experience of the problems that can arise when criminal liability is used in this manner for the enforcement of regulatory legislation in situations of the type described. Norway, as a country that has industrialised more recently, has come later to these same problems. The report which preceded the new Norwegian legislation noted that concern had been expressed by some that imposing strict liability on corporations for offences under petroleum laws would be contrary to the Norwegian philosophy that no one should incur criminal sanctions unless he was at fault. It was suggested that criminal liability might be imposed on corporations for 'cumulative errors'. Cumulative errors would be present when a violation was the consequence of several errors committed by different persons within the company, where the errors committed by individual persons were not sufficiently serious to warrant criminal proceedings. It was, however, decided that this would be to impose too strict a liability on the corporation. Therefore it was decided to impose criminal liability on the corporation only where it could be seen that one person had been at fault (although it would not be necessary to identify the person); the company would not be liable for the acts of an individual which were committed in 'disloyalty' to the company. It was considered that to impose liability on the corporation would create an incentive to management and other responsible parties to avoid violations of the law.[1]

It is significant to observe that, in both Britain and Norway, although their criminal laws have developed from very different legal traditions, similar solutions are being arrived at in enforcing regulatory standards offshore. It is not, therefore, unreasonable to identify a process of 'convergence' between the jurisprudence of the two countries in the enforcement of offshore regulatory legislation against what are very similar problems.

Corporate and personal responsibilities under offshore legislation

The enforcement of standards offshore by means of legal sanctions is dependent on the creation of a realistic pattern of legal responsibilities. This necessity applies equally to civil and criminal liabilities.

British Legislation

Under the Petroleum (Production) Regulations, systematic attempts have been made to attach a series of safety provisions to petroleum licences; however, speaking in legal terms, the effectiveness of these attempts depends in the last analysis upon the questions: 1) Upon whom do they place responsibilities?; and 2) If they are not observed, upon whom may sanctions be imposed?

The essentially 'personal' nature of the petroleum licence conditions ensures that responsibilities can only be imposed upon the licensee; if these are not fulfilled, sanctions can thus only be imposed upon the licensee. Any other entity or person will be a stranger to the contract between the Department and the licensee.

In reality, the licensee will certainly delegate many of the functions and responsibilities under his petroleum licence to others, who may be his employees, contractors, and the employees of others. However the legal principle will be inviolate: one may delegate performance of the licence contractually to others, but one may not delegate legal responsibility; in the event of failure to meet the conditions of the licence, only the licence holder will be legally liable.

Where attempts have been made to enforce standards offshore through criminal sanctions, the legislature has taken a much more sophisticated approach to the linked problems of responsibility and culpability.

Petroleum legislation

The Mineral Workings Act makes an attempt to attach responsibility for the norms that it seeks to enforce to the entities and persons most likely in practice to deal with the matters in question. Matters[2] concerning the organisation of the activity are seen as the primary responsibility of the concession owner in relation to the licence; obligations relating to structural and equipment matters are made the primary responsibility of the owner of the installation; while matters concerning the on-the-spot running of the installation on a day-to-day basis devolve upon the installation manager.

In order to clarify the exact limits of responsibility in complex situations, the law makes all three prima facie responsible, until such time as the parties who had no responsibility – or who had discharged what responsibility they did have

- have established these facts and thus exculpated themselves under the legislation.[3]

As in the case of the petroleum licence, delegation beyond that provided for in the Mineral Workings Act, is not an acceptable defence to either responsibility or criminal liability under the Act. The owner of an offshore installation will be held responsible for failure to ensure that there are in existence written instructions for the purpose of Reg 7 of the Operational Safety, Health and Welfare Regulations, whereas the offshore installation manager is likely to be held responsible for failure to carry them out. Similarly, under Reg 5 there has to be in force a scheme which, amongst other matters, requires the examination, and where necessary the testing, of equipment within specified intervals. The owner of the installation is responsible for ensuring there is such a scheme, while the offshore installation manager will be responsible for any failure to have it carried out. If the work is contracted out, the maintenance contractor will have no criminal responsibility for his failure to maintain the machinery beyond the somewhat limited duties set out below, whatever may be his civil liability for a breach of the maintenance contract.

Under Regulation 32(3) of the same Regulations –
It shall be the duty of every person while on or near an offshore installation –
(a) not to do anything likely to endanger the safety or health of himself or other persons on or near the installation or to render unsafe any equipment used on or near it;
(b) to co-operate with his employer, if employed, and any other person upon whom a duty or requirement is imposed by these Regulations so far as is necessary to enable that duty or requirement to be performed or complied with; and
(c) to report immediately to the appropriate responsible person, or if no such person be appointed, or if appointed, unavailable, to the installation manager, any defect in any equipment which appears to him likely to endanger the safety, health or welfare of persons on or near the installation or the safety of the installation any equipment used with it.

It is significant that, apart from the obvious disciplinary requirements, these Regulations relate to co-operation with employers or 'persons' upon whom duties are imposed by the Regulations: as has been mentioned previously, in Reg 5, a maintenance obligation is imposed upon the installation owner and manager. It is arguable that a contractor or his workmen to whom the maintenance work had been delegated would be guilty of an offence under Reg 32(3)(b) if, through his neglect in carrying out maintenance work, Reg 5 were to be contravened.

The Act and Regulations impose few responsibilities, other than those discussed above, upon the persons employed offshore by the concession owner or installation owner, and fewer still upon sub-contractors or their employees. Apart from the Diving Operations at Work Regulations,[4] it is difficult to identify direct legal obligations imposed upon a sub-contractor offshore.

Where direct legal responsibilities are imposed upon 'any person' offshore, a duty is placed upon that person's employer by Reg 32(2) of the same Operational Safety, Health and Welfare Regulations to ensure that he complies with that obligation.

In fact, the main thrust of the mineral workings legislation is to place the primary day-to-day responsibility offshore upon the installation manager. For this purpose he is given quite remarkable powers of direction and control over all persons on or near his installation, which go beyond anything otherwise currently encountered in British employment law. Thus, under S5 of the Act he is given powers of direction over employees both of his own employer, and of sub-contractors, including the right to detain for limited periods; furthermore, wilful disobedience to his commands by any person subject to his authority constitutes a criminal offence.[5] It seems to follow from S5 that an employee of a sub-contractor, although he may have no direct criminal responsibility, for example, for the safeguarding of machinery which he may bring with him on to an installation, would commit an offence if he disobeyed an installation manager's instruction to remove the machine, or at any rate, not to use it in its unsafe condition.

The criminal liability of installation and concession owners under the legislation will be tempered by their undoubted corporate status. Corporate bodies may be convicted under British criminal law, but in practice only financial penalties can be imposed upon them. However, the Mineral Workings Act contains a clause – common in onshore legislation – providing for the separate conviction, in his individual capacity, of any 'director, manager, secretary, or other officer' who may be said to have contributed to the law breaking of his corporation by his individual conduct: ie a concept of 'reverse' vicarious liability.[6]

Responsibilities under the Petroleum and Submarine Pipe-lines Act[7] may be placed upon 'persons'. In most instances this is likely to be the pipe-line owner, or proposed owner with a similar clause for individual responsibility as that referred to in the previous paragraph.

The Health and Safety at Work, etc Act

With the application offshore in 1977 of the Health and Safety at Work, etc Act, the pattern of duties, responsibilities and criminal liabilities became more complex. First, the Act applies not only to the installation owner or oil field operator, but to all employers of labour or sub-contractors, large or small, whose activities relate to an installation, or the area around it.[8] Second, the Act imposes novel responsibilities upon employers and management for the safety of those who are not their employees; a vital consideration in the crowded conditions prevalent offshore,[9] where one man's safety may depend upon the precautions taken by the employees of another. Third, the Act imposes duties upon every employee while at work – irrespective of seniority or function – to take reasonable care for the safety and health of himself and all other persons who may be affected by his acts or omissions at work, and to 'co-operate with his employer or any other person' to enable those persons to comply with their duties under the Act.[10] These provisions, as interpreted by the courts, have imposed a 'management-style' responsibility upon all who, in the course of their work, have to discharge duties delegated to them by their seniors in the chain of command.

It should be borne in mind that the general duties under the Act must be complied with to the extent that is reasonably practicable. What is reasonably practicable for a particular manager to achieve will depend upon his status within the organisation, his powers and his control over resources. For instance, if a new process requires safeguards costing £10,000, it would certainly be deemed reasonably practicable for an employer to provide them. But if it is sought to impose the obligation to provide them upon a manager, would he be responsible if his employer had not provided the £10,000? It may be that a court would hold that a manager could not be responsible for a matter where he lacked the financial 'power to prevent' the contravention which lay outside his span of control; but specific provision may be needed in the legislation for the protection of managers in a situation such as this.

Although the Act makes no specific reference to offshore installation managers, it appears clear, from the nature of the powers and duties bestowed upon them, both in practice and under the mineral workings legislation, that they will also bear a heavy management-style responsibility under the Health and Safety at Work, etc Act. Although primary duties are imposed upon employers, S37 of the Act imposes liability upon any 'director, manager, secretary or other officer' whose consent, connivance, or neglect has caused, or contributed to, the offence in question.

A Director of Roads of a large local authority was, in the unreported case of *Armour* v *Skyne*, convicted under S37 following a fatal accident which was deemed to have occurred as a consequence of the director's failure to implement the local authority's safety policy, in contravention of the duty which the policy itself imposed upon him.

Although it might be argued that this 'boardroom level' responsibility is inappropriate to describe the ambit of the large, but necessarily geographically concentrated, powers and responsibilities of an offshore installation manager, the latter may well be caught by S36 of the Act, which provides that where the commission of an offence under the Act by any person (such as an employer) is due to the act or default of some other person, that other person shall be guilty of the offence. It is easy to imagine situations offshore where the respective conduct of employer and offshore installation manager could be brought within the ambit of this section.

In any event, an offshore installation manager could clearly be caught by S7 of the Act, which although directed at employees generally has been interpreted by the courts so as to impose legal responsibilities proportional to the span of control of those in management positions: it is difficult to imagine an 'employee' under S7 who would carry a heavier responsibility for compliance with the Health and Safety at, etc Work Act than the installation manager.[11]

Thus, the managerial pattern of responsibility, which is arguably present under the mineral workings legislation, emerges with full clarity from the Health and Safety at Work, etc Act in its application offshore. For the first time there is legislative provision for a system that embraces the complexities of the multi-employer situations that exist on and about an installation.

The general duty in S2(2)(a) to ensure safety and health through the provision and maintenance of 'systems of work' introduced a new dimension into employers' responsibilities offshore. It extended quite dramatically the concept of written instructions as required by Reg 7 and the Second Schedule of the Operational Safety, Health and Welfare Regulations. This concept is based upon case law developed by the civil courts in employers' liability cases.

As originally formulated,[12] an employers' duty with respect to safe systems of work was four-fold: to provide the necessary equipment, to provide the necessary trained and skilled workers to use it, to organise the work in advance, and lastly, to take steps to ensure that the laid-down system was followed. As it has been developed more recently, an employer's duty is seen as 'assessment and response' to a hazard. First, an employer is under a duty to assess the risks of his operation; he must then respond positively by planning a safe system adequate to meet the hazard, including training and information for his employees. He must also ensure, so far as is reasonably practicable, that the laid down system is followed; and lastly, he must monitor his own performance, to ensure that the scheme is adhered to, and that the hazard has been minimised.[13]

In certain instances, a further duty arises to report to a regulatory authority on the results of the monitoring process. It is rare to identify this last stage in British offshore legislation, but more common in Norway. It is significant that Regulations embodying this latest development of the safe systems approach are now being made, extending S2(2)(a) to particular hazards.[14] In the case of the Lead at Work Regulations 1980, those already apply offshore, and the proposed Control of Substances Hazardous to Health Regulations, could be extended offshore when they are made.

Perhaps the most relevant aspect of these new-style Regulations is that they contemplate a situation in which an employer may be responsible, and criminally liable, if he fails to enforce his own laid-down system against his employees or, possibly, the employees of others. In the latter situation, it appears from the cases that in most instances, a duty to inform rather than to enforce will be the more readily found by the courts.[15]

An excellent example of the 'safe systems' approach will be found in the Ionising Radiation Regulations 1985, made under the Health and Safety at Work, etc Act, and applied offshore. No employer shall carry out work with ionising radiation until he has assessed the hazards and taken reasonably practicable steps to minimise those hazards within the parameters of the Regulations, which include local rules, contingency plans, monitoring of exposure, dose limitation and notification of occurrences.

The British Merchant Shipping Acts

These Acts originally imposed most of their responsibilities upon the captains of British-registered ships with, for example, certain structural and stability responsibilities upon the owners. The increasing tendency in recent years for this legislation to regulate for health, safety and welfare on British ships has meant that more responsibility and criminal liability is being imposed upon the captain, or 'persons',[16] on broadly the same lines as that imposed upon the installation manager, or other persons, under the Mineral Workings Act.

Norwegian legislation

The considerations and limitations discussed in relation to British petroleum licences might be said to apply equally to Norwegian licences; particularly the need to impose obligations upon parties other than the contracting parties. However, from the beginning the Norwegian legislature introduced a 'hybrid' statutory form of licensing, which has succeeded in overcoming many of the shortcomings inherent in a 'pure' licence system.

Responsibilities under the Norwegian petroleum licence

The responsibility for compliance with the conditions of a petroleum licence is imposed upon the licence holder under the Petroleum Act of 22 March 1985.[17] The licensee will, as in the British system, normally engage others to perform responsibilities and functions under his licence. These other persons may be employees, sub-contractors, or the employees of others. At this level the licensee, as in Britain, will remain responsible for the failure of performance of his delegate, but at that point the similarity between the two systems comes to an end.

Where the two systems differ is in the impact of the Regulations deemed to be incorporated into the licence under S57 of the Act. These may impose duties upon third parties as well as the licensee, and result in either or both of two kinds of sanction similar to, in English law: 1) a 'civil' type sanction directly related to the licence, and imposed upon the licensee; and 2) a 'criminal' type sanction directed against whoever may be the party responsible.

These two concepts are reinforced by the fundamental duty imposed upon the licensee under S58 of the Petroleum Act to 'ensure' that anyone working for him either personally or through employees, contractors or sub-contractors shall comply with the licence and any legislation associated with it (which in practice will be all the regulatory legislation with which the licensee must himself comply).

This concept, known in Norway as 'internal control', is developed further in the Royal Decree of 28 June 1985,[18] which, in S4, places a similar duty of enforcement upon 'anyone conducting or taking part in petroleum activities'. While the licensee has the main responsibility for enforcement, the section also places a duty upon every contractor, large or small, to whom any of the licensee's duties have been delegated, not only to comply personally but also to enforce compliance upon their own employees; taken literally, it also imposes duties upon the individual workmen to carry out their work in accordance with the Regulations applicable.

The Norwegian authorities clearly attach great importance to this principle of internal enforcement, as exemplified by the detail contained in the separate Royal Decree – also of 28 June 1985 – in which the systems required for internal control are set out. These systems, which are similar to, but a great deal more elaborate than, the concepts of 'safe systems of working' recently developed in Britain, show the same reliance upon assessment of risks, and structured response. The Norwegian system differs in the main in the documentation that must be preserved.

The question remains whether the responsible person would still be criminally liable under these provisions if he had done his best to enforce the Regulations, but without success; in other words, would his criminal responsibility, when the Regulation in question was contravened by an employee or agent, be personal or vicarious? In Norwegian criminal law, since the emphasis is upon personal criminal liability, it appears unlikely that a licensee or contractor would be held liable, regardless of his personal fault, under a provision of this type. Offences might be committed both by 'perpetrators and non-enforcers', but in the latter case a defence comparable to that of 'due diligence' (familiar in British regulatory legislation) might be admitted.

Thus the responsibilities imposed under the Norwegian licence system can best be seen as an amalgam of the British licence system, together with the pattern of responsibility under the British Mineral Workings Act. The result may be conveniently summarised as follows:

In the event of a breach of, for example, legislative provisions concerning safeguarding of machinery on an installation[19] there would be a breach of the licence by the licence holder.[20] The following consequences might occur:

1. The civil sanction of forfeiture of licence.[21]
2. Prosecution of licensee for failure to comply with, or to enforce the terms attached to the licence.[22]
3. A criminal offence committed by a contractor or sub-contractor to whom the duty to safeguard machinery had been delegated.[23]
4. Possibly a criminal offence committed by the actual offender whose negligence or other breach of duty was the actual cause of the contravention.[24]

The criminal sanctions imposed under the Petroleum Act may be either fines or imprisonment, which in 'aggravated circumstances' may be for up to two years. As licences will be held by 'companies or other associations', human individuals will not be personally liable as licensees: thus, in practice fines may only be imposed upon the licence holders in cases of breaches of the licence conditions or Regulations. However, it seems clear that individuals may also be convicted under S66 of the Petroleum Act for breaches of the Act or of particular Regulations that may be applicable. Moreover, the penal provisions of the 1985 Royal Decree relating to safety are expressed impersonally, so that either corporate licence holders or individual managers, contractors or workers might be prosecuted for a breach of these Regulations.

Section 66 of the Petroleum Act also makes provision for the vicarious liability of the 'legal persons' (ie licence holder or contractor) for offences committed by anyone acting 'on their behalf', with heavier penalties in those cases where the contravention has been committed to promote the interests of the legal person, and where that person has in fact benefited from the contravention.

It is interesting to observe that, in this complex new machinery for imposing legal responsibility upon the licensee and the hierarchy of managers, contractors and employees through whom offshore activities are carried out, there is no concept of the 'qualified strict liability' with which students of British regulatory legislation are familiar. By this it is meant that, under the British system as applied offshore, a series of named persons are deemed to be responsible for a failure to observe the Regulations, but, if prosecuted, may individually exculpate themselves by proof of satisfactory conduct.[25] No such defence is available under the Norwegian Petroleum Act, Royal Decrees or Regulations, but under the Norwegian system individuals would not be liable unless the prosecution established their personal wrongdoing.

Responsibility under the Working Environment Act

The application offshore of the Norwegian Working Environment Act adds a further dimension to the patterns of responsibility already discussed.

The Act regulates the activities of employers and employees, as defined in Ss3 and 4. The term employer, includes, for purposes of responsibility, persons conducting the enterprise in the employers' stead (ie members of senior management). The general duties of the employer are set out in S14, in terms reminiscent of S2 of the British Health and Safety at Work, etc Act. The duties of employees are in S16; again these resemble the employees' duties in S7 of the Health and Safety at Work, etc Act, but they are set out in greater detail and include a special duty imposed upon employees who are leaders or supervisors to ensure that safety and health are taken into consideration when work is carried out. Duties are also laid upon manufacturers and suppliers of technical apparatus, equipment, and toxic and other noxious substances, again on the lines of S6 of the Health and Safety at Work, etc Act.

A failure to comply with these duties on the part of any proprietor, employer, or person managing in the employer's stead, committed wilfully or through negligence, can result in fine or imprisonment under S85. Employees may also be criminally liable for 'complicity' and subject, under S86, to a fine only. Where 'aggravating circumstances' are present, the penalty under S85 may be increased to up to two years' imprisonment, although clearly this increased penalty may only be inflicted on individual, as opposed to corporate, defendants.[26] Aggravating circumstances include risk to life or health, or evidence of failure to comply with instructions from the enforcement agencies, or requests from employee safety representatives or safety committees.

In the case of offences involving risks to health and safety, employers or their managers may escape the heavier penalty where it can be shown that they have acted in a 'fully satisfactory manner'; but the Act does not make clear upon whom is the onus of establishing this defence. The tradition of Norwegian criminal law would suggest that the burden lay on the prosecution to establish fault.

As already indicated, employees who through negligence commit a breach of the Act or Regulations are subject to a fine under S86, but where the offence is committed wilfully or through gross negligence, the penalty may be up to three months' imprisonment. In very aggravated circumstances, the penalty may be up to one year. Gross negligence is not defined, but aggravating circumstances include cases where special safety directives were not observed, or where the employee knew, or should have known, that the offence could have seriously endangered the life or health of others.

Provision is made in S87 for the vicarious liability of a corporate employer for a failure of a person acting on its behalf to comply with an enforcement notice issued by the enforcing agency. As in the case of S66 of the Petroleum Act, the financial penalty that may be imposed under this section will be increased where the offence was committed to promote the interests of the enterprise, or if the enterprise did in fact benefit therefrom.

Responsibility under Norwegian flag legislation

Mobile drilling units and other mobile installations which are registered in the Norwegian register of ships are subject, on a world-wide basis, to the standards of construction, operation and manning enforced by the Maritime Directorate. In the past, a reading of the legislation and Regulations suggested that little attention had been paid to the issues discussed in this chapter: thus, apart from occasional references to the duty of the owner and platform manager to enforce the Regulations, the only criminal responsibility was that contained in the Norwegian Penal Code, so that any contravention of the regulation in question was punishable under either Ss339(2) or 416 of the Code. In the present system, both these sections, which make it a criminal offence to fail to comply with a statutory Regulation, are still couched in terms of an essentially personal criminal responsibility:

> Fines may be imposed upon anybody who ... violates any regulation ... (S339(2))
> Any shipmaster or owner who fails to observe the duties which according to Norwegian law are incumbent on him ... (S416)

In addition, however, Regulations dated 28 June 1985 impose obligations on the owners of Norwegian mobile units with regard to internal control, on the same lines as those discussed as part of the petroleum legislation.

Where, however, a Norwegian-registered mobile unit is operating on the Norwegian continental shelf, it also becomes subject to the petroleum legislation already discussed, and to the pattern of responsibilities set out in the Petroleum Act.

References

1   *Odelsting Proposition No 72* Ministry of Petroleum and Energy, 1982-83, pp63-65.
2   Mineral Workings (Offshore Installations) Act 1971 S5.
3   Ibid, S9(3).
4   Reg 8.
5   Mineral Workings (Offshore Installations) Act 1971 S5(3).
6   Ibid, S9(1).
7   Petroleum and Submarine Pipe-lines Act 1975 S28.
8   Health and Safety at Work, etc Act S2(1).
9   Ibid, Ss3 and 4.
10   Ibid S7.

11 See Chapter 7 for further consideration of the importance of the Health and Safty at Work, etc Act offshore.

12 See *Wilsons and Clyde Coal Co Ltd v English* [1938] AC 57.

13 See, eg the Lead at Work Regulations 1980.

14 Control of Industrial Major Hazards Regulations 1984.

15 *R v Swan Hunter Shipbuilders Ltd* [1981]. See Chapter 7.

16 Merchant Shipping Act 1970 S19 – criminal liability of master, owner, and other persons. Also the Merchant Shipping (Health and Safety: General Duties) Regulations 1984.

17 Ss56-58.

18 Reg 3 – 'The Licensee shall through internal control ensure that the activity is in accordance with provisions stipulated in, and in accordance with, the acts set out in Reg 1' (ie the Petroleum Act and Royal Decrees).

19 Worker Protection and Working Environment Act of 4 February 1977.

20 Act of 22 March Ss56, 57, 58.

21 Ibid, S62.

22 Ibid, Ss58 and 66.

23 Royal Decree of 28 June 1985 concerning safety. Ss4 and 59.

24 Ibid. S59.

25 See Mineral Workings (Offshore Installations) Act 1971 S9(3).

26 S51 of the Petroleum Act empowers inspectors to issue 'orders'; but is silent as to the sanctions available to secure compliance – hence the significance of S85 of the Worker Protection and Working Environment Act.

5

The Enforcement of Offshore Safety Legislation

Principal legislative provisions

British legislation

Merchant Shipping Acts 1894–1985.
Continental Shelf Act 1964.
Mineral Workings (Offshore Installations) Act 1971.
Health and Safety at Work, etc Act 1974.
Petroleum and Submarine Pipe-lines Act 1975.
The Offshore Installations (Inspectors and Casualties) Regulations 1973.
The Offshore Installations (Operational Safety, Health and Welfare) Regulations 1976.
The Submarine Pipe-lines (Inspectors) Regulations 1977.
The Diving Operations at Work Regulations 1981.

Norwegian legislation

Worker Protection and Working Environment Act 1977.
Act pertaining to petroleum activity 1985.
Regulations for mobile drilling units 1973 and other Regulations and Guidance† noted in Appendices 1 and 2 within Chapter 6.
Regulations concerning exploration and drilling made by Royal Decree of 29 August 1975.*
Regulations concerning safe practices for production made by Royal Decree of 9 July 1976.*
Regulations concerning safe operations in petroleum activity made by Royal Decree of 28 June 1985.
Regulations concerning the licensee's internal control made by Royal Decree of 28 June 1985.
Norwegian Penal Code

The nature of the problem

Legislation is only effective when it is enforced. This is as true for offshore safety legislation as any onshore legislation; but the problems of legal enforcement of safety standards offshore are of a different order from those with which onshore enforcement agencies are familiar. The principal problems that have

---

* Largely obsolete since Act of 22 March 1985.
† See note to Chapter 4.

emerged – leaving aside the extremely hazardous nature of the activity itself – are as follows:

(1) The problem of jurisdiction.
(2) The advanced nature of the technology (comparable to that of the onshore nuclear power industry, for example) to be regulated.
(3) The integrity standards required for a harsh environment.
(4) The identification of the standards to be enforced.
(5) The constantly shifting nature of the workforce (comparable to the construction industry onshore) to be regulated.
(6) The interface between industrial and maritime risks and responses.
(7) Finally, a problem that may be peculiar to the industry: the very different degree of development of the countries upon whose continental shelves the petroleum activity takes place. As a consequence, the operations may perhaps be supervised by coastal jurisdictions employing sophisticated governmental agencies possessing considerable enforcement powers, and fully experienced with regulating advanced technology; or, perhaps, by jurisdictions in which regulatory agencies and statutory standards are either rudimentary or unknown.

While some of these factors are to be found in industry onshore, considering all of them together offshore, it is not surprising that offshore enforcement of safety standards presents no simple coherent pattern (in contrast, for example, with the world-wide enforcement of road traffic standards by local police), but is a patchwork of differing and potentially overlapping, or even conflicting, requirements of the various public agencies.

Nevertheless, despite the differences, certain fundamental safety considerations are common to both the onshore and offshore working environments. Plant and equipment must be both provided and maintained to a high standard. Manpower must be carefully selected. The standard of qualifications for key personnel must be high. Training must be appropriate and comprehensive. Systems of work must be exact, well conceived and rigidly imposed.

Standards will differ from onshore, however, particularly as regards the integrity of the installation and its plant and equipment in relation to the harsh environment. There will be specialised areas of training in survival, well control, fire fighting and emergency procedures on which there is superimposed the constraint of installation size.

Selection of personnel by medicals taking account of not only physical, but also mental fitness is of importance for personnel, both in respect of the offshore isolation and the work patterns which have developed. To these difficulties is added that of the installation/shipping interface.

Jurisdiction remains perhaps the most difficult enforcement problem of all. Oil activity may take place in the open sea, involving mobile installations, ships and fixed production platforms, with large numbers of persons at risk. A coastal state will license the operator to engage in the activity. The operator, if he owns an installation, may directly employ labour in his activities, and employ contractors and their equipment and employees. Mobile installations will probably be hired with their crews, as will shipping; on fixed installations there is likely to be much sub-contracting by the owner to other organisations, who will have their own employees. All feasible legal control of these activities may conveniently be classified as the formulation and enforcement of shelf or petroleum standards (discussed in Chapter 2); licence standards (Chapter 2); flag or shipping standards (Chapter 3); and Certification standards (Chapter 6).[1]

'Shelf' standards refer to those standards which are enforced under the jurisdiction of the coastal state. The state concerned applies either specific offshore legislation, or part, or even the whole, of its onshore legislation to its continental shelf, enforceable against any persons, irrespective of nationality, engaging in petroleum activity therein.[2] The application of safety standards offshore by the coastal state, and the enforcement of these standards by civil and criminal sanctions in the onshore courts, are then relatively straightforward.

'Licence' standards are a relation of shelf standards and in fact may pre-date them. A licence, being essentially 'personal', creates no problems of jurisdiction; it may be made in one place, and observed, or broken, anywhere in the world. Theoretically, any breach of any clause can result in the sanction of forfeiture, but it is unrealistic to bank upon forfeiture occurring over a safety issue. There is no other legal sanction, apart from that provided for in the terms of the licence, for the breach of a 'pure' licence; and, by definition, no means of enforcement against any person other than one of the parties to the licence.

It may be that the coastal state has neither technical expertise, legal resources nor inclination to interfere in this way, so that the organisations engaging in petroleum activity on its continental shelf may find themselves very largely unregulated as far as safety is concerned – apart from any flag standards that may be applicable to their ships, or structures treated as ships under the shipping legislation.[3]

A mobile installation is likely to be controlled to some extent by the flag standards of its country of registration. Under shipping law, it will either be subject to the national legal standards concerning structure, seaworthiness, and manning applied by that country to ships generally; or, in the case of some maritime nations (eg Norway, USA), it will be subject to specific standards for mobile drilling platforms or units.[4]

It should be noted that flag states differ greatly in the extent to which they legislate for ships in their registry: some content themselves with seeking the approval of a certifying authority (discussed below) while others, such as Great Britain, tend to regulate structure, performance, system and employment conditions on ships flying their flag to an extent analogous to onshore employment Regulations.

Flag standards may be enforced in the courts of the flag state, and inspections by that state may be carried out either in port or elsewhere (British Department of Transport inspectors are empowered to board and inspect British ships wherever they may be).[5] The essential point is that flag standards operate at times and locations when the vessel may also be subject to the jurisdiction of a 'shelf' state. Potential conflicts of jurisdiction and standards are reconcilable on the basis that the mobile installations concerned must always meet the higher standard in such a case, be it flag or shelf.

Certification standards (also known as classification standards) are another importation from the world of shipping. These developed from the practice of London maritime insurers in the 18th century, who would rely upon an assessment and certificate from an independent source as to seaworthiness in fixing insurance premiums for a particular ship. This practice became general, and a number of recognised certification, or classification, societies have emerged. The essential feature is the contractual nature of the process: a ship owner for his own purposes employs a Society to check upon the construction and equipment standards of a ship, as a protection both at the construction and commissioning stage.

Although intended solely for the protection of the ship owner, the certification or classification came to be relied upon by third parties as a standard of construction and equipment – eventually, such classifications came to play a role in safety. Nowadays, certain flag jurisdictions rely upon the 'class' of a ship in accepting it for registration under their flag. This process has been extended to mobile units as vessels. It has also been incorporated into the approvals system operated by some offshore jurisdictions, so that an installation either fixed or mobile – will not be accepted on to their particular continental shelf without certification by a recognised certifying authority as to its equipment and structure.

The certification procedure, as an enforcement technique, is somewhat analogous to, but may be less effective than, the licence procedure. Since it is a contract between two private parties, it can only be enforced by them, even though it may be relied upon by third parties, including the state. It may contain ongoing stipulations relating to maintenance and replacement of structures or equipment as a condition for remaining in classification; but, unless the certification authority interests itself in and enforces these clauses, and generally monitors their observance, the existence of the certificate may generate a feeling of false confidence. Both the Alexander Kielland[6] and the Ocean Ranger[7] disasters throw light on this aspect of the certification/classification system.

British legislation

Petroleum legislation

Licence standards

Section 1 of the Continental Shelf Act, which extended the UK state licensing system offshore, made provision for 'model clauses'[8] in any licence granted for offshore exploration or production. These model clauses were carried forward to the Petroleum (Production) Regulations 1982. They included a requirement that the licensee should comply with any instructions given by the Secretary of State for Energy in writing for securing the safety, health and welfare of persons employed in or about the licensed area.[9] Similar considerations can relate to the authorisation of a pipe-line under the Petroleum and Submarine Pipe-lines Act 1975 (infra).

Procedures for the enforcement of petroleum licences will be found within the terms of the particular licence. The standard licence permits persons authorised by the Department of Energy to enter any of the licensee's installations, to inspect installations, wells, plant, appliances, and works; and where unsafe conditions are found, to execute any appropriate remedial action and recover the cost. Revocation of the licence is a possibility if there is any breach of the terms of the licence, which could include any written safety clauses.

In theory, enforcement of a clause attached to a licence should be simple within contractual norms, the sanction being either damages or revocation of the contract, ie the licence. But this deterrent for enforcement of safety clauses in licences does not, on the face of it, seem to be credible – there are no records of revocation of a licence on the United Kingdom continental shelf on these grounds.

A further weakness in licensing is that, being personal, it imposes no obligations upon third parties, although the terms of the licence will commonly make the licensee responsible for the conduct of his sub-contractors, such as drilling companies or catering organisations. The licensee will, in his turn, only be able to exercise any degree of control over his agents, contractors or employees whose acts or omissions may endanger his licence, through the mechanism of his contracts with them.

The enforcement of the safety provisions of a British petroleum licence may be summed up as follows:

1. The licence purports to control all the petroleum activity of the licensee within that part of a designated area on the British sector of the continental shelf to which the licence relates.
2. Health and safety instructions may only be issued to the licensee, but their scope may not necessarily be limited to the safety and health of persons in the direct employment of the licensee.
3. Powers of entry and inspection under the licence appear to be wide; indeed there is nothing which expressly confines them to fixed or mobile installations.
4. It is unclear whether these licensing provisions provide a mechanism for the control of activities (such as diving) carried out on the licensee's behalf by independent contractors; but it is difficult to see how a licensee could resist a requirement making him answerable to the Department of Energy for the conduct – or, for that matter, the safety and health – of the employees of any diving contractor he might employ.
5. The sanctions for breach of safety and health instructions are either revocation of the licence, or, arguably, a further instruction ordering the cessation of the activity in question until the matter is resolved. Alternatively, the Secretary of State might himself abate the hazard and require the licensee to reimburse the cost; however, the writers can cite no example of this procedure having been employed.
6. The contractual nature of the licence precludes the imposition of sanctions in respect of unsafe practices upon any person other than the licensee.

Continental shelf standards
The Mineral Workings (Offshore Installations) Act 1971 was the first sys-

tematic attempt to formulate specific health, safety and welfare standards for offshore operations, and to enforce them legally through criminal sanctions.

The Act imposes limited duties, and criminal responsibility for non-compliance, upon the owner of the installation, the concession owner and the installation manager.[10] It is significant that this legislation makes little distinction between drilling and production installations, fixed or mobile.[11] In this respect the British system diverged from the path taken by other states with an interest in offshore activity.

Responsibility for enforcement of the Act and Regulations devolves upon the Department of Energy, who have appointed inspectors for this purpose.

The powers of inspectors appointed under the Act are set out in the Inspectors and Casualties Regulations. These are broadly similar to those granted to inspectors under British onshore safety legislation. Their powers are confined to installations: inspectors may board and inspect an installation, or the adjacent sea-bed. To exercise these powers the inspector may require the owner or manager of an installation to convey him to or from the installation at any reasonable time,[12] also to provide him with reasonable accommodation and subsistence while on board.

The inspectors' right of access and inspection extends to any 'installation' whether or not it is capable of being manned.[13] An inspector having gained access to an offshore installation has the right to: gain access to all parts; inspect the installation and its equipment; inspect and take copies of any certificate of insurance, entry in the logbook or any other document or record relating to the installation and its operation. He can test, or in certain cases remove or dismantle equipment, and can require the manager or any other person to do, or refrain from, any act that appears necessary or expedient for the purposes of averting a casualty or minimising the consequences of one. However, this latter action can only be imposed in the first instance for a maximum of 72 hours.

Fatalities and casualties or other accidents involving danger to life suffered by a person on or near an installation have to be notified immediately to the owner by the installation manager, and in turn the owner shall as quickly as possible notify the Department of Energy. In the meantime the manager has to submit a written report to the owner who, within three days following receipt of this information, has to notify certain particulars of the casualty in writing to the Department of Energy.

Currently other accidents, injuries or diseases suffered by a person on or working from an installation, or an attendant vessel, in an operation on or in connection with the installation, which cause a person to be disabled from work for a continuous period of more than three days have to be notified once a quarter to the Department of Energy.

Thus an inspector visits an installation for two main reasons: routine inspection and the investigation of a casualty. He will on occasion make other visits, in particular to follow up on matters resulting from a routine inspection and to deal with a particular matter of specific importance.

The inspectors' enforcement powers contained in the Inspectors and Casualties Regulations are related to three main criteria:

1. A contravention of the Mineral Workings (Offshore Installations) Act 1971 and its Regulations.
2. Absence of or breaches of any conditions on certificates or examination reports required to be carried out under the various sets of Regulations.
3. A hazardous situation which may or may not also represent a contravention.

In the event of a contravention, the inspector has several options: he may agree remedial action with the manager, entering matters requiring attention in the logbook; he may write to the owner outlining matters of importance and requesting appropriate corrective action; or he may consider the matter serious enough to initiate a prosecution.

Breaches of the Act or Regulations are treated as criminal offences punishable summarily or upon indictment, in the onshore courts, with maximum penalties of an unlimited fine and/or imprisonment for conviction upon indictment.[14] As mentioned previously, where a hazard appears to involve the risk of a 'casualty', defined as involving a risk of loss of or danger to life, the inspector is empowered to require the owner, manager, or any person on board or near to an installation to take action to avert the casualty. Compliance with the instruction may involve stoppage of all or part of the operations on the installation, or an activity (eg diving) being carried on near to it. Such orders cannot remain in force for more than 72 hours without confirmation by the Secretary of State; the installation owner is entitled to make representations to the Secretary of State before the latter confirms or quashes the inspector's requirement.

It will be recalled that the Department of Energy administers the petroleum licence, as well as the mineral workings legislation; and that the former entitles the Secretary of State to issue specific safety instructions. However, this licence procedure imposes direct pressure upon the licensee only, as contrasted with the mineral workings legislation, which imposes sanctions – and therefore pressure – upon a variety of named persons.

There is no explicit link between the petroleum licence procedure and the enforcement of the Mineral Workings Act: thus breaches of the licence remain essentially 'contractual', while breaches of the Act give rise to criminal sanctions; the only link being that certain hazardous situations may lead to a breach of the licence as well as giving rise to contravention of the Act. This lack of formal automatic linkage between breaches of a petroleum licence and breaches of the criminal law offshore represents a major difference between the British system and that of Norway.

The Petroleum and Submarine Pipe-lines Act 1975
This Act makes provision for Regulations relating to the safety and health of persons employed on pipe-lines and pipe-line works, and for criminal penalties for breaches of these Regulations.

Failure to comply with the Act or Regulations constitutes a criminal offence.[15] Where pipe-line works are carried out in contravention of the Act or Regulations, the Secretary of State may serve a notice upon the person concerned, requiring him to remove the works, or to take such steps as the

Secretary of State considers necessary to render them safe.[16] It is interesting to note that to date no Regulations specifically relating to the health and safety of persons, other than for diving and the reporting of accidents, have been made under this Act.

Enforcement of the Act is assigned to the Secretary of State for Energy, who has appointed inspectors for this purpose.

The powers and functions of inspectors in enforcement of the Pipe-lines legislation are set out in the Submarine Pipe-lines (Inspectors and Casualties) Regulations. Inspectors are empowered to enter any premises, vessels, or installation,[17] 'used or intended to be used in connection with a pipe-line', and also enjoy a power – similar to that granted to inspectors under the Mineral Workings Act – to demand the provision of free transportation and accommodation offshore.

Inspectors may inspect as a matter of routine, or investigate accidents which have occurred relating to a pipe-line, or in pipe-line works, which occurrences must be reported to the Department of Energy in the same way as for installations. These inspections will often, though not exclusively, be conducted on vessels from which the pipe-line works are conducted, rather than from installations as this term is normally understood.

Legal responsibilities under the Act and Regulations are imposed upon 'persons';[18] in most instances this will be the pipe-line owner or proposed owner.

Where contraventions of the Submarine Pipe-lines Act are discovered, the inspectors may take action similar to that described for installations. Again, if punitive action is taken it would be to:

1. Initiate a prosecution. Contraventions of the law are criminal offences, triable either summarily or on indictment in onshore courts. Penalties are the same as those provided in the Mineral Workings Act.[19]
2. Issue a 'requirement', under the Inspectors Regulations.[20] These requirements are similar to those under the Mineral Workings Act.

Health and Safety at Work, etc Act 1974
This legislation differs from that discussed above in that it is not specific to offshore situations, but is general onshore legislation that has been applied offshore. The broad objectives of this legislation include securing the health, safety, and welfare of persons at work, and protecting persons other than persons at work against risks to their health and safety arising out of, or in connection with, the activities of persons at work. It imposes wide general duties upon employers, employees, contractors, manufacturers or suppliers of articles and substances for use at work, and controllers of premises. These duties may be expanded by the very comprehensive regulation-making powers contained in the Act.

The Act and Regulations are enforced on land by inspectors employed by the Health and Safety Executive. Non-compliance constitutes a criminal offence,[21] with a range of penalties similar in extent to those already referred to under the mineral workings and submarine pipe-lines legislation, but with additional punitive powers given to inspectors.

However, enforcement offshore is by inspectors appointed by the Department of Energy under an agency agreement[22] made between the Secretary of State and the Health and Safety Commission. Under this agreement the Department's inspectors may exercise all the powers granted to the Executive's inspectors under the 1974 Act: which powers, in addition to the powers to enter, to inspect, to receive information, to prosecute and to investigate accidents and dangerous occurrences, entitle inspectors exercising powers under the 1974 Act to issue improvement and prohibition notices, and to take statements from relevant persons by direct questioning.

Improvement notices[23] may be issued in writing by an inspector to 'any person' who, in the opinion of the inspector, is, or has, contravened any of the Relevant Statutory Provisions (in this instance, the General Duties of the Act and any of the Regulations that are applicable offshore) requiring that person to remedy the contravention within a stated period.

Prohibition notices[24] may be issued by an inspector to any person carrying on, or in control of, activities which, in the opinion of the inspector, involve a risk of serious personal injury. The prohibition notice follows the same procedure as the improvement notice, with the significant difference that the inspector's direction that the specified matters be remedied is accompanied by an order that the activity in question shall cease until this is done: either forthwith, (when they are known as immediate prohibition notices) or after a specified period, (deferred notices). An appeal against the issue of either an improvement or a prohibition notice may be heard by an industrial tribunal.[25] Failure to observe the terms of an inspector's notice is a criminal offence, which can be punishable, in the case of a prohibition notice, by up to two years' imprisonment, following trial by indictment.[26]

It is noteworthy that the United States equivalent legislation, the Federal Occupational Safety and Health Act 1970, has been given specific offshore effect. S4(a) states:

> This Act shall apply with respect to employment performed in a State, ... and outer continental shelf lands as defined in the Outer Continental Shelf Act.

Standards under this Act, including record keeping and reporting of accidents, have been applied to offshore activities: by a Memorandum of Understanding (equivalent to a British Agency Agreement) enforcement powers offshore have been delegated to the US Coast Guard, making the latter the 'lead enforcement agency' offshore.

Other legislation
There remains some uncertainty as to the full range of legislation that can be enforced offshore. Apart from the specific range of legislation that was so applied by the Continental Shelf Act,[27] it is difficult to decide with precision what legislation is, or is not, enforceable offshore. When Acts are specifically applied, eg Radioactive Substances Act by virtue of the Continental Shelf Act, there remains uncertainty as to whether the powers of their respective enforcement agencies can be exercised offshore. In the event, no attempt has been made by

the agencies concerned to function offshore, so that the question of the enforcement of this legislation remains uncertain.

It must be assumed that the criminal laws protecting the person and property apply offshore; at least since the enactment of the Police and Criminal Evidence Act 1985[28] it is intended that the police may have the right of access to offshore installations, but the extent of the British criminal law that they will be able to enforce thereon has yet to be clarified.

It may be that in many instances the only sanction for unlawful behaviour is that an operator could invoke contractual terms against an offending employee; remove him from the installation, discipline or dismiss him.

Flag standards: the enforcement of the Merchant Shipping Acts

The contribution of shipping law to offshore safety, health and welfare was assessed in Chapter 3. For present purposes it should be appreciated that supply, stand-by, pipe-laying, and other vessels taking part in offshore activity may be registered in a country that has no interest in the offshore operation, but which may have legislated for the safety, health and welfare of the crew of that ship, with world-wide effect. It is also likely that a mobile drilling unit will itself be registered as a ship, with the concomitant obligation to observe such of the flag state's legislation as relates to drilling vessels.

In the case of Great Britain, much legislation relates to crew conditions and working hours, but there has been nevertheless, a tendency in recent years to legislate and formulate Regulations applicable to the safety, health and welfare of the crews of British-registered ships on a world-wide basis.[29] The issue may be complicated by the dual role – maritime and industrial – of a mobile drilling rig;[30] for when it is on station, the maritime crew will be augmented by a drilling crew, who will all be bound by the drilling safety Regulations enforced by the shelf state. (For the British sector of the North Sea, these will be found in Schedules 2 and 4 of the Operational Safety, Health and Welfare Regulations, applicable to any drilling activity taking place on the British continental shelf.) Moreover, the drilling consents applicable by the Petroleum (Production) Regulations 1982 will also deal directly with safety.

The Merchant Shipping Acts 1894-1984, together with supporting Regulations, provide a complex code covering construction, seaworthiness, manning, crew conditions, health and safety at work and related issues on British registered ships. This legislation is enforced by the Department of Transport. Activities in the North Sea over which the Department currently exercises supervision include British supply vessels, British stand-by vessels, and the small number of mobile drilling units which have been registered as British ships under the Merchant Shipping Acts. Potential areas of supervision might include:

1. Any crane barge constructing fixed platforms that might – in the future – be British-registered.
2. Any British-registered accommodation vessel in the vicinity of, or perhaps attached to, an installation.

3. Any British-registered barge engaged in the construction or maintenance of a pipe-line on the British continental shelf.

The Department of Transport normally enforces this legislation on the basis of port inspections but, since 1979, its inspectors may board and inspect a British ship anywhere in the world. Failure to comply with the provisions of the Merchant Shipping Acts and Regulations is punishable by prosecution: either before magistrates, with financial penalties which may be imposed upon owners, masters, and other persons; or on indictment, when penalties may be an unlimited fine or up to two years' imprisonment.[31]

Since 1984, Department of Transport inspectors have been empowered to issue improvement and prohibition notices, on much the same basis as those issued under the Health and Safety at Work Act.[32] Notices may be issued with regard to activities 'on board any ship'. It is significant that prohibition notices may relate to serious personal injury to any person 'whether on board ship or not',[33] which creates the possibility of the issue of a prohibition notice in interface situations involving ship/installation, ship/construction, or ship/demolition activities.

The thrust of a notice is to stop an 'activity' or to prevent a ship from going to sea. Appeal from a Merchant Shipping Act notice lies to a single arbitrator, as opposed to the appeal to an industrial tribunal provided for under the Health and Safety at Work Act.[34] Professional qualifications, as specified in S4(5), are required for an arbitrator. An important difference between the two systems is that, if an arbitrator finds that a prohibition notice has been invalidly served, he may award compensation, payable by the Department of Transport to the ship owner.[35] There is no equivalent provision under the Health and Safety at Work Act. Penalties for failure to comply with an inspector's notice are on the standard Merchant Shipping Acts' scale, with a maximum of two years' imprisonment for non-observance of a prohibition notice.

Department of Transport activity overlaps with offshore activity in a number of ways, including:

1. Where a mobile installation is registered as a British ship, it will have to comply with the appropriate merchant shipping legislation, as well as the appropriate continental shelf legislation. A distinction is made, however, between 'shipping' and 'petroleum' activities. For example, neither the Merchant Shipping (Health and Safety: General Duties) Regulations, nor the Merchant Shipping (Safety Officials and Reporting of Accidents and Dangerous Occurrences) Regulations apply to a mobile installation on its working station; nor is an accident reportable under the latter Regulations where it is reportable under the Offshore Installations (Inspectors and Casualties) Regulations.

2. The control of support, supply, stand-by and any pipe-laying vessels which might be British-registered. These must comply with the Merchant Shipping code; but a stand-by vessel of any nationality should satisfy the guidance issued by the Department of Transport, in order that the

installation which it serves may be regarded as in compliance with the offshore installations legislation.

3. The inspection and certification of fire-fighting and life-saving equipment on all offshore installations in controlled waters has been made the responsibility of the Department of Transport inspectors, because of their experience in enforcing the like standards on British ships over many years.[36]

Criminal proceedings

Where proceedings have been instituted for offences committed offshore, there appears to have been a preference for reliance upon the Health and Safety at Work, etc Act. The bulk of all prosecutions taken since 1977 have been brought under that Act. While most of these have charged breaches of S2, a significant number have been brought under S3, indicating, it might be said, the growing importance of the system's approach to offshore safety.[36a] Overwhelmingly, these prosecutions have been brought before courts of summary jurisdiction: resulting, where guilt has been established, in fines of, on average, well below £1,000. Three cases taken on indictment (ie before juries) resulted in fines of £5,000, £10,000 and £15,000. There is no record of any sentence of imprisonment having been inflicted in any of these proceedings.

From the nature of the offences charged under the Health and Safety at Work, etc Act, it appears that most prosecutions have been brought against employers. There is no information as to how many of the charges brought under the mineral workings legislation were brought against installation managers[36b] or other officials, either alone, or together with an installation or concession owner. Nor is there a record of any 'petroleum-related' prosecution brought by the Department of Transport under the merchant shipping legislation.

Norwegian legislation

The first licences on the Norwegian continental shelf were granted in 1965.[37] As the system developed,[38] specific conditions could be applied to them. Regulations, linked to the subject matter of the licences, could be issued by the King.[39] Provision was made for the appointment of inspectors to enforce both the licence and all the conditions attached to it.[40]

The licensee was declared responsible for the observance of all the conditions of his licence by himself, his contractors and their sub-contractors.[41] As in the British licensing system, 'serious or repeated' violations of the provisions of a licence entitled the Minister to revoke the licence.[42]

More detailed safety regulations were issued in the Royal Decrees of 25 August 1967, 3 October 1975 (Exploration) and 9 July 1976 (Production). Again, the standards formulated by these Decrees formed part of any licences granted under the Act of 1963; anyone carrying out the licensed activity was bound to comply with the applicable safety codes contained in the Decrees, and to ensure compliance by others.

Breaches of these requirements empowered the Minister to revoke the licence[43] and – in sharp contrast to the British system – the same breaches were also deemed to be[44] criminal offences punishable under the Penal Code S339. The cumulative effect of these concepts was to render a licence holder criminally liable not only for his personal failure to observe the Regulation in question, but also for his own failure to enforce it against others; also to hold the actual offender, whoever he might be, criminally liable.

Thus, the significant ways in which the Norwegian licence enforcement system differed from the British were:

1. The Norwegian effort to avoid the purely 'contractual' weaknesses exemplified in the British licensing system.
2. The imposition of criminal sanctions for breaches of the licence conditions, which arguably introduced a new dimension into the offshore licence.
3. The enforcement of these criminal sanctions against both licensees and third parties.[45]

Evidence of the success of the Norwegian experiments may be found in the influence of these characteristics on the new Petroleum Act of 1985, which now makes provision for the granting of exploration and production licences. The Act continues to make provisions for Regulations linked to licences.[46] Section 50 makes provision for 'suspension of the activities', and S62 makes provision for revocation of the licence for serious or repeated violations of laws, Regulations, stipulations or conditions related to the licence. 'Wilful or negligent' violations of the licence conditions are declared to be criminal offences, punishable by fines or imprisonment for up to two years.[47]

Section 58 re-enacts the duty of the licensee to ensure compliance by his delegates, in much the same terms as those contained in the superseded Royal Decrees.

A number of Royal Decrees have been issued under the Petroleum Act, the most significant for safety being the two issued on 28 June 1985.[48] The first concerns safe operations in petroleum activity on the Norwegian continental shelf. This Decree has two main aspects: 1) an application system for all phases of petroleum activities; and 2) the functions necessary for the performance of prudent operations required by S45 of the Petroleum Act. It thus sets out a framework for safe operations, on the basis of the primary responsibility of the licensee for his own activities, and for the activities of those working for him, directly or indirectly. This system of responsibility, known as 'internal control', is supervised and enforced by the Petroleum Directorate which, under the new legislation, assumes primary responsibility for the enforcement of safety standards offshore. Again, breaches of these Royal Decrees are treated as 'civil' breaches that might lead to a loss of licence under S62, or as a criminal offence under S66.

The second Royal Decree of the same date sets out in detail the concept of the licensee's duty to exercise 'internal control'. It sets out the requirements for the operator's administrative system for internal control to enable compliance with the safety requirements of the legislation.

With the coming into force of the Act and the Royal Decrees on 1 July 1985, the earlier framework legislation was repealed, except: first, a substantial corpus of Regulations will remain in force until they are progressively assimilated into the new system; and second, the former Regulations may still apply in so far as they are incorporated in earlier licences.

Functions of the Norwegian enforcement agencies

Introduction

Historically, the number of Norwegian government agencies with offshore enforcement potential has been remarkable: although the system has been much modified since the legislation of 1985, the legacy of the previous system is still apparent.

In the former system, the emphasis on the difference between fixed and mobile installations, and the division of responsibilities – with the Maritime Directorate effectively responsible for mobile installations, and the Petroleum Directorate effectively responsible for production installations – served to confuse the distinction between the petroleum and the shipping legislation in their application to installations.

In the former system, distinctions had in some instances to be made between the agency which initiated legislation, and the Agency in practice responsible for its enforcement offshore. Thus, although the Maritime Directorate was directly responsible for co-ordinating approvals for mobile installations, other agencies (such as, for example, the Aviation Administration dealing with use of helicopters) not only set standards, but were also involved in the process of approvals. Similarly, agencies other than the Petroleum Directorate set standards for production platforms. In the case of mobiles, the Seamen's Directorate had some responsibility for labour conditions aboard Norwegian platforms: on production platforms the Petroleum Directorate organised labour conditions with delegated powers from the Ministry of Labour under the Working Environment Act.

For ease of comprehension, some reference must still be made to the former system, before the changes introduced by the Petroleum Act can be fully assimilated.

The control of mobile installations under the former system

The original basis of control was known as the 'two-track system': safety Regulations made under the continental shelf legislation were applicable to all mobile drilling units operating on the Norwegian continental shelf, irrespective of nationality; concurrently, safety Regulations made under Norwegian shipping legislation were applicable to mobile platforms of Norwegian registration world-wide.

The continental shelf safety legislation involved contributions from a number of Norwegian government agencies, although the co-ordinating of the activities of these agencies was carried out by the Maritime Directorate, which also gave the final consent to the use of a drilling platform on the continental shelf. The enforcement system was a combination of approvals and routine inspection: for example, the NPD were responsible for the approval of the drilling equipment on the mobile under evaluation, and also issued the vital drilling permit for the commencement of drilling on the continental shelf.

The agencies concerned, together with their areas of responsibility, were as follows:

The Maritime Directorate: design and strength, stability, emergency services and equipment, helicopter decks, work inspection, including cranes, ladders, safety equipment, work procedures.

The Petroleum Directorate: technical drilling equipment, drilling procedures, control of divers, and emergency preparedness plans.

The Aviation Administration: helicopter decks, including location, design, and equipment.

The Telecommunications Directorate: location of radio room, radio and radiotelephone equipment, and other telecommunications equipment.

The Directorate of Seamen: organised safety work and registration of persons on board.

The Health Directorate: medical office with equipment, first aid equipment, hygienic conditions, medicines, doctor system, qualifications of first-aiders.

Waterfalls and Electricity Administration: electrical facilities and area classification for explosion hazard, zones and ventilation, electrical installations and equipment in dangerous areas.

Institute for Radiation Hygiene: transportation, storage and use of radioactive equipment.

Explosives Inspection: storage and use of explosives.

The Maritime Directorate also carried out the necessary inspection on behalf of the Institute for Radiation and the Explosives Department.

Norwegian-registered mobiles were regulated mainly by the Seaworthiness Act and Regulations; also to a lesser extent by the Seamen's Act. The agencies which took regulatory responsibility for Norwegian mobiles were:

The Maritime Directorate
The Waterfalls and Electricity Administration
The Telecommunications Directorate
The Petroleum Directorate
The Classification Societies.

In the case of Norwegian mobiles also, the Maritime Directorate acted as the co-ordinating agency, and granted the final permission for the use of the mobile installation.

The current control system for mobiles under Norwegian law
The system as described above was simplified, at least as far as the number of agencies involved offshore was concerned, by the passing of the Petroleum Act in 1985 and the coming into force of the various Royal Decrees made under it.

The new Act regulates all petroleum activities of exploration, drilling, production, utilisation, and pipe-line transportation on the Norwegian continental shelf as well as in internal and territorial waters. It is specifically applied to installations and vessels being used in the listed activities, but does not apply to the movement of installations.

Authority for the enforcement of the Act and the associated Royal Decrees, as well as for the applicable provisions of the Working Environment Act, against all mobile installations operating on the Norwegian continental shelf, has been delegated to the Petroleum Directorate, together with the responsibility for co-ordinating the activities of such other agencies retaining enforcement powers in these matters on the continental shelf (eg, Ministry of Social Affairs relating to hygiene, and Ministry of Justice for public rescue services).

Details concerning this delegation of authority will be found in the Royal Decree of 28 June 1985 (supervisory activities) from which the dominant role of the Petroleum Directorate will be discerned. In discharging its new responsibilities, the Petroleum Directorate may seek expert assistance from the following agencies:

The Telecommunications Administration
The Civil Aviation Administration
The Explosives Directorate
The Coast Directorate
The Meteorological Institute
The Maritime Directorate.

Where mobile installations of Norwegian registry are concerned, the Maritime Directorate remains responsible, as before, for regulating compliance with the Seaworthiness Act and Regulations and other shipping legislation. A Norwegian mobile platform newly coming on to the Norwegian continental shelf will thus become subject initially to the jurisdiction of the Maritime Directorate and, on entry to the shelf, to the Petroleum Directorate; however, in practice, the standards under both sets of safety legislation are drafted and enforced so that conflicts of standards will be unlikely, even though it would clearly be the duty of the licensee to ensure that the mobile installation he was utilising met the standards of the shelf upon which he was operating.

The control of fixed platforms: the former system
The departure point for the control of fixed platforms on the Norwegian continental shelf was the licensee's general responsibility for safety, expressed to a considerable extent through the internal control system functioning within his organisation. Thus, the duty of the licensee was both to comply with the applicable safety Regulations, and to ensure that the Regulations were complied with within his organisation by means of an internal control system. Regulations to this effect were made under both the continental shelf legislation and the Working Environment Act. The enforcement agencies responsible, under this body of legislation for fixed platforms, together with their related areas of activity, were:

The Petroleum Directorate: control of platform layout with concept evaluation, including load-carrying structures, pipe-lines, shipment facilities, drilling equipment and procedures, pressure vessels and pressurised systems, machinery and auxilliary equipment, electrical facilities, area classification, fire prevention, gas warning equipment, fire warning equipment, fire extinguishing, emergency shut-down systems, emergency power generation, emergency lighting, alarms, internal communication, cranes, ladders and handrails, living quarters, diving operations, diving equipment, qualification requirements and emergency preparedness plans.

The Maritime Directorate: rescue equipment and its location, launching equipment for lifeboats, rescue exercises.
The Telecommunications Administration: Telecommunications systems.
The Coast Directorate: marking and identification.
The Aviation Administration: aviation operation conditions.
The Ministry of Health and Social Affairs: medical examination of personnel, control of substances hazardous to health, medical office, ventilation, lighting and furnishing of the living quarters.
The Ministry of Justice: emergency preparedness which has or which may lead to, loss of human lives and which requires immediate evacuation.

The co-ordinating responsibility was imposed upon the Petroleum Directorate.

The present system
The principal responsibility for the enforcement of the Petroleum Act and Royal Decrees on fixed platforms under Norwegian jurisdiction is imposed upon the Petroleum Directorate by the Royal Decree of 28 June 1985 (supervisory activities). This Directorate also co-ordinates the activities of such other agencies as retain enforcement powers in that area (eg the Ministry of Justice relating to emergency preparedness). To discharge this function the Petroleum Directorate may seek the expert assistance of the following agencies:

The Telecommunications Administration
The Civil Aviation Administration
The Directorate of Fire and Explosion Prevention
The Coast Directorate
The Meteorological Institute
The Maritime Directorate.

The primary objective of the Petroleum Directorate in fulfilling its new objectives in the case of fixed installations will be the monitoring of the systems for internal control in petroleum activities on the Norwegian continental shelf, as set out in detail in the Royal Decree of 22 March 1985. Thus it appears that the former system of regular inspection as a tool of enforcement will now be confined to the control of Norwegian mobiles by the Maritime Directorate. In both the areas of Petroleum Directorate responsibility, namely fixed and mobile installations operating upon the Norwegian continental shelf, the main thrust will be to monitor the way in which the licensee discharges his self-regulating duties through his organised systems, organised and observed in the shadow of the licence; and the serious consequences to him that any loss of the licence, for this or other reason, would entail.

Powers of the Norwegian enforcement agencies

Under the 1985 Petroleum Act, there can be no doubt that the principal enforcement powers of the enforcement agencies relate to the petroleum licence. The function of the Petroleum Directorate under S51 of the Act is to monitor the observance of the exploration and production licences, with all their associated terms and conditions, and the many and varied Regulations which are deemed, under Ss5 and 8, to form part of the licence. For this purpose the inspectorate may issue orders, and grant consents and approvals under S56; these powers are set out in detail in the Royal Decree of 28 June 1985. In the event of breach, the Petroleum Directorate has the following powers to secure compliance:

1. Issue a directive under Ss50 and 51, which may be enforced by coercive fines levied under S65. Further details concerning these fines will be found in the Royal Decree dated 3 June 1977.
2. Revoke the licence in cases involving serious or repeated violations of the terms of the licence or Regulations (S62).
3. Withdraw or suspend the consent granted for the particular phase of the operations. Details concerning consents will be found in the Royal Decree dated 8 June 1985 (safety). Consents are discussed in Chapter 6.
4. Lay criminal charges under S66 for 'wilful or negligent' violations of the Act or Regulations, including the terms of the licence. Punishment is by fines or imprisonment for up to three months; in particularly aggravating circumstances, the imprisonment may be for up to two years. Although it appears that the Act contemplates a personal criminal liability in the first instance, the section goes on to make provision for the vicarious criminal liability of the organisation (legal person) on whose behalf the convicted person was acting, with a remarkable provision for the penalty to be increased where the organisation had 'benefited' from the criminal offence.
5. Expel offending ships or aircraft from the Norwegian continental shelf; or alternatively, seize them and bring them to a Norwegian port (S65).

In the enforcement of the Worker Protection and Working Environment Act offshore, the inspectors can rely upon the specific enforcement provisions incorporated in that Act. The powers and duties of the inspectors are contained in Chapter 12 of the Act. Apart from the standard Norwegian system of 'consents and approvals', the inspectors may, in appropriate cases, issue orders enforced by coercive fines under S78. It appears that the 'orders' issued under S77 are of two kinds. First, general orders, the non-compliance with which is punishable by coercive fines levied under S78. Second, orders issued where there is a danger to life or health: these must be obeyed forthwith, on pain of the inspector exercising his power to close down the enterprise, or the relevant part of it.

Any manager or employer who wilfully or through negligence commits a breach of the Act, or any order issued under it, commits a criminal offence, punishable under S85 by a fine, or three months' imprisonment, or both. Again, in the event of 'very aggravating circumstances' the penalty may be up to two years' imprisonment. The like penalties may be imposed upon any employee under S86.

References

1 For an excellent critique of flag, shelf and approval standards, see *Report of the Royal Commission on the Ocean Ranger Marine Disaster* Canada, 1984, pp3–7.
2 See Norwegian Petroleum Act of 22 March 1985, S2.
3 For shipping legislation generally, see Chapter 3.
4 Norway: see Regulations set out in the *Red Book*; US see Department of Transportation *Requirements for Mobile Offshore Drilling Units* 1979, 46 CFR Chap 1–A; Canada see *Recommendations 13 and 14 of the Ocean Ranger Report*.
5 Merchant Shipping Act 1979 S27.
6 See *Report No 67 to the Storting, on the Alexander L Kielland Accident* Chapter 8. This was the worst industrial disaster in Norwegian history, involving loss of 123 lives.
7 *Ocean Ranger Report* Chapter 1, pp3–7.
8 Continental Shelf Act 1964 S1(4).
9 Schedule 5, clause 23.
10 Mineral Workings (Offshore Installations) Act 1971 S3(4).
11 Ibid, but see S7(7): different Regulations may be made for installations which are registered vessels, or in transit.
12 The Department does not operate its own transport system.
13 SI 1973, No 1842, Reg 1(2).
14 Mineral Workings (Offshore Installations) Act 1971 S7(3).
15 Petroleum and Submarine Pipe-lines Act 1975 S28(1).
16 Ibid, S28(2).
17 SI 1977 No 835.
18 Petroleum and Submarine Pipe-lines Act 1975 S28.
19 Ibid, S29.
20 The Submarine Pipe-lines (Inspectors, etc) Regulations, Reg 3(1)(d).
21 Health and Safety at Work, etc Act 1974 S33.
22 Agency agreement made between the Health and Safety Commission and the Secretary of State for Energy, taking effect on 1 November 1978.
23 Health and Safety at Work, etc Act S21.
24 Ibid, S22.
25 Ibid, S24.
26 Ibid, S33(4)(ed).
27 Continental Shelf Act 1964 S4, applies the Coast Protection Act 1949; S6 applies the Wireless Telegraphy Act 1949; S7 applies the Radioactive Substances Act 1960, and S8 the Telegraph Act 1885.
28 Police and Criminal Evidence Act 1985 S23, but not yet in force.
29 Merchant Shipping Act 1979 S21: Merchant Shipping (Health and Safety: General Duties) Regulations 1984.
30 Recognised by the IMO as 'a special purpose ship designed and operated to carry out an industrial function at sea'.
31 Merchant Shipping Act 1979 Ss21 and 22.
32 Merchant Shipping Act 1894 S1.
33 Ibid, S2.

34 Ibid, S4.
35 Ibid, S6.
36 The Offshore Installations (Fire-fighting Equipment) Regulations 1978; the Offshore Installations (Life-saving Appliances) Regulations 1977.
36a Discussed post pp121–123.
36b An installation manager was connected in 1987.
37 Under the Act of 21 June 1963.
38 Formalised under Royal Decree of 8 December 1972.
39 Ibid, S38.
40 Ibid, S45.
41 Ibid, S54.
42 Ibid, S57.
43 Exploration S119; Production S129.
44 Production S62; Exploration S120.
45 S63.
46 S57.
47 S66.
48 1. Regulations Concerning Safety in Petroleum Activities; 2. Regulations Concerning the Licensee's Internal Control in Petroleum Activities.

Part II

Safe Systems and Damage Control

6

Installations, Licences, Surveys and Approvals

Principal legislative provisions

British legislation

Petroleum (Production) Act 1934.
Continental Shelf Act 1964.
Continental Shelf (Designation of Areas) Orders 1964.
Mineral Workings (Offshore Installations) Act 1971.
The Offshore Installations (Registration) Regulations 1972.
The Offshore Installations (Managers) Regulations 1971.
The Offshore Installations (Logbooks and Registration of Death) Regulations 1972.
The Offshore Installations (Construction and Survey) Regulations 1974.
The Petroleum (Production) Regulations 1982.

Norwegian legislation

Act of 22 March 1985 No 11 pertaining to petroleum activities.
Regulations supplementing the act pertaining to petroleum activities made by Royal Decree of 14 June 1985.
Regulations concerning safety, etc relating to the Act concerning petroleum activities made by Royal Decree of 28 June 1985.
Regulations concerning the licensee's internal control in petroleum activities on the Norwegian continental shelf made by Royal Decree of 28 June 1985.
Regulations for mobile drilling platforms† laid down by the Maritime Directorate on 10 September 1973 and all other Regulations and guidelines listed in Appendices 1 and 2 within this chapter.

The activities of exploration for and exploitation of petroleum resources on the continental shelf of the North Sea ('petroleum activities') cannot be lawfully undertaken except under a licence granted by the coastal state within whose sector of the shelf the activity is to be conducted. Petroleum activities may be

---

† See note to Chapter 4 p68

undertaken by the licensee personally or by an organisation operating under a concession ('a concession owner') granted by the licensee.

In the Norwegian and British sectors of the North Sea, an organisation operating within the authority of a licence is required to satisfy the coastal state as to the suitability of the installation or installations which it intends to employ in its petroleum activities. That is to say, the coastal state must be satisfied that the installation is structurally sound for the purpose and for the place where it is to be used. The structure's suitability will be evaluated in conjunction with the system under which it is to be operated.

The system of operation laid down in the operations' manual will be considered fully in subsequent chapters concerned with safe systems of work and emergency procedures.[1] In this chapter the emphasis will be on the regulatory provisions – largely concerned with documentation – which must be complied with before petroleum activities can commence.

In the Norwegian sector start-up procedures relate also, and independently, to conceptual, construction and operational phases of the offshore activity: theirs is a 'stop-go' system under which the licensee is 'stopped' before the commencement of each phase of his activity and required to obtained further consents.

Licensing

Licensing is primarily concerned with the grant of the right to exploit the submarine resources coupled with arrangements for taxing the petroleum recovered. Nevertheless, the licensing procedure does present an opportunity to control the organisations to whom these valuable rights are granted – an organisation with a poor safety record could, for example, be refused a licence.

The licence also provides a framework for control of the petroleum activities themselves. In the Norwegian system the emphasis throughout the regulatory codes is on the licensee's responsibilities. The tight system of narrowly defined duties, which are the consequence of the British system having been based in the criminal law, result in the licensee being a less obvious candidate for responsibility in some situations in that system, but nevertheless it is common for British Regulations to place responsibilities on the licensee as concession owner, as well as the owner of the installation and the installation manager. In the early days of offshore development, before the regulatory codes had been enacted, it was common for British licences to impose general safety responsibilities on the licensee.

British legislation

The principal licensing provisions in the British sector stem from the Petroleum (Production) Act of 1934 which was originally intended to apply only to exploration for and exploitation of resources on the mainland of Great Britain. That Act vested the property in petroleum existing in its natural condition in strata in Great Britain in the Crown and gave the Crown the exclusive right to search and bore for and get such petroleum.[2] It also provided, however, for the granting to 'such persons as they think fit licences to search and bore for and get petroleum'.[3]

The Continental Shelf Act 1964 vested any rights exercisable by the United Kingdom outside territorial waters with respect to the sea-bed and subsoil and their natural resources in the Crown and extended the 1934 Act to enable granting of licences in respect of such resources. It also enabled model clauses placed in licenses to include provision for the safety, health and welfare of persons employed on operations undertaken under the authority of any licence.[4]

The Petroleum (Production) Regulations 1982 made detailed provision for model clauses in production licences according to whether production was to occur in 'landward' or 'seaward' areas as these expressions were defined within the Regulations.

The United Kingdom has not purported to exercise even the limited sovereignty which the Geneva Convention allows to coastal states, over the whole of its sector of the North Sea continental shelf. Consequently, licences have not been granted until the area which is to be opened for petroleum activities has been 'designated' as an area to which British jurisdiction will be extended as far as is necessary for the purposes of these activities.[5]

Norwegian legislation

Section 3 of the Act of 22 March 1985 pertaining to petroleum activities asserts the State's right to subsea petroleum resources. Section 5 gives to the Ministry the authority to grant a licence to explore for petroleum, the licence to be for a period not in excess of three years. Section 4 stipulates that the licence shall designate the areas it applies to.

Section 8 gives the power to grant production licences to the King in Council, and outlines the procedure which will be followed for advertising the intention of the state to grant production licences. When a production licence is granted, S8 requires that the Ministry shall appoint an operator to conduct the day-to-day management of the petroleum activities in accordance with the licence. Section 45 contains the general requirement:

> Activities pursuant to this Act shall be conducted in a prudent manner and shall take due account of the safety of personnel and environment ...

Regulations supplementary to the Act of 22 March made by Royal Decree of 14 June 1985, stipulate the procedure to be followed by the organisation wishing to apply for either an exploration[6] or a production[7] licence. Section 47 states that the Ministry may require that a licensee plan and organise a control and documentation system for ensuring compliance with requirements laid down in or pursuant to the Act or these Regulations.

Regulations concerning safety made by Royal Decree of 28 June 1985 contain the main consents which have to be obtained, and the conditions which have to be complied with in order to obtain these consents, before the licensee may engage in petroleum activities.

Surveys and approvals

Apart from satisfying all the criteria for a licence, the first major safety problem affecting any licensee/owner is that of meeting the requirements of the regulatory authority in respect of the offshore installation, whether it is to be used for drilling or production or a combination of both. On the other hand, the sheer bulk and complexity of an installation presents the regulatory agency concerned with the same difficulty as that facing any agency concerning itself with sophisticated and detailed technology: that of identifying and utilising the necessary expertise so as to, at least, match the expertise available to the 'client' organisation. This problem has hitherto been tackled to some extent by conventional regulatory techniques, and to some extent by involving independent expertise through a system of certification by independent experts.

At the outset the Norwegian regulatory agencies played a more active role in this exercise than did the British. Under the new Norwegian regulatory system, with its emphasis upon the licensee's responsibility for internal control, it is arguable that the situation has been reversed and in the Norwegian sector more reliance is placed upon the provision of expertise by the organisation engaged in the petroleum activities than is now the position in the British sector.

British legislation

Instead of laying down detailed safety criteria for an installation, and seeking by means of systematic inspection and supervision to be satisfied that the installation, with all its complexities, meets those standards, the British government has followed the long-standing practice of shipping insurance, at least as far as structural safety and equipment are concerned.

The law now, in these respects, merely lays down certain safety criteria which must be met: proof of compliance with these criteria is furnished by means of a certificate from one of the certification authorities, which have been appointed for this purpose. Five of the six so appointed are ship-certifying authorities who have a world-wide reputation in this role. If one of these authorities is prepared to certify the completed, or modified installation, the Department of Energy is satisfied.

Thus the Mineral Workings (Offshore Installations) Act 1971, S3, together with the Offshore Installations (Construction and Survey) Regulations 1974, require installations installed or working on the British sector of the North Sea continental shelf to be of the appropriate structural standards, and properly equipped for the conditions in which they will operate. A current certificate of fitness testifying to these standards is required of the owners of both fixed or mobile installations constructed upon, or brought within, the British sector of the continental shelf.

This certificate refers to the total structure, and also its equipment, and the circumstances in which it will be used.[8] The climatic conditions in which it is intended to operate are clearly a matter of importance and, for example, mobile drilling units intended to operate in far northern waters will not be acceptable unless they have been constructed to the standards necessary to enable them to meet the climatic challenges of those waters.[9]

The criteria with regard to which the certification authority must be satisfied are set out in Schedule 2 of the 1974 Regulations. These are based upon internationally recognised standards for offshore oil practice, and include environmental considerations, foundations, primary and secondary structures, fittings, materials, construction and equipment. This certificate must be commissioned by the owner of the installation who must acquire it from a recognised certifying authority and pay the cost of the necessary supervision and survey by that authority. The certifying authority must be approved by the Secretary of State.[10]

The certificate, once granted, is valid for the period specified, which may not exceed five years from the last major survey carried out by the certifying authority. In addition, the Department of Energy issues guidance to the certifying authorities on these matters in what is known throughout the industry as the *Blue Book*. The guidance is continually updated and modified and periodically revised. Alterations, or damage to an installation may well require a further survey and any failure to comply with requests could lead to the certificate being revoked by the Secretary of State for Energy.

It follows that the authority, in carrying out a survey, acts as agent of the commissioning owner, and in no way binds the enforcing authority by the standards which it adopts. Thus, in the (unlikely) event of an omission in a survey carried out by a certifying authority, an installation owner who acted on the strength of the survey and put the installation into operation on the British continental shelf might still be deemed to be in contravention of the Act and Regulations, and be prosecuted. In practice, however, the Secretary of State might be more likely to censor or even withdraw his recognition from the certifying authority that was at fault, and might terminate the certificate of fitness.

The important point is that the certifying authority does not, by making a survey, bind the enforcing authority; the law merely makes the issue of a satisfactory certificate to the owner in respect of the installation a condition precedent to its operation on the British continental shelf.

It will be appreciated that the British 'approval' or 'certification' standard is intended to operate, in the case of mobile installations only, on a shelf-wide basis, but may have world-wide implications. A fixed installation, of course, will be used in one location only; but a mobile installation, having secured its certificate from a world recognised certification organisation, may be able to satisfy the construction requirements of most, if not all, the offshore jurisdictions subject to any differences in climate and other environmental requirements. An exception would be an offshore jurisdiction that does not follow the common practice of international recognition of the certification authorities. In the event of an offshore jurisdiction, having higher or differing standards, a society or organisation could be involved to survey the installation at these higher standards.

In this respect the certificate of fitness for installations, at least in British practice, has something in common with the world-wide recognition of flag standards for ships operating on the high seas away from the legal jurisdictions of any power other than the 'flag' power which the ship owner obeys. Such flag standards in many instances nowadays include construction and equipment standards, which the ship owner must meet if he wishes to continue to receive the protection of the – say – British flag on the high seas. In the same manner a mobile drilling unit may be constructed in a Japanese shipyard to the standards of a certain classification society, but operate throughout its life in the Atlantic basin; what will be controlling upon it will be not the construction standards imposed by domestic Japanese legislation, but those certified by the classification society, which will be recognised world-wide. More so if there is incorporated into the construction the requirement of the coastal state as regards certification, under which control it will, or is likely to operate.

In the British sector construction standards are laid down by the standards of the Construction and Survey Regulations and the Guidance for both fixed and mobile installations.

Where mobile installations are registered as British ships they will in addition, need to comply with any relevant structural or equipment standards in force under the Merchant Shipping Acts.[11] In the case of foreign mobiles operating on the British continental shelf, it will be appreciated that, in addition to compliance with British certification and shelf standards, they may carry with them on to the British continental shelf some flag standard requirement imposed by the country of registration whose laws follow them world-wide – including on to foreign continental shelves.[12]

Even with a valid certificate of fitness, before the installation is operated in the British sector it must be registered under the Offshore Installations (Registration) Regulations 1972, Reg 4 of which requires:

(a) no fixed installation shall be established in the relevant waters;
(b) no mobile installation shall be brought into those waters with a view to its being stationed there, and
(c) no fixed or mobile installation shall be maintained in those waters, unless it is registered.[13]

Regulation 5 requires that for the first registration of an offshore installation the following particulars must be included:

(a) the name and address of the person or persons seeking to register it;
(b) where no address furnished pursuant to head (a) is an address in the United Kingdom, an address in the United Kingdom to which communications for the owner may be sent;
(c) a name or other designation for the installation;
(d) particulars of any other registration of the installation (whether as a vessel or otherwise and whether in the United Kingdom or elsewhere);
(e) an indication of the nature and the function or proposed function of the installation;
(f) an indication whether the installation is a mobile or a fixed installation and, in the case of a mobile installation, whether it has its own motive power;
(g) if the application relates to a part of an installation, particulars of any major additions to be made;

(h) an indication of the location at which the installation is stationed or intended to be stationed in the relevant waters;

(i) an indication of the period for which it is expected the installation will be stationed at the location;

(j) in the case of a mobile installation, its tonnage.

The owner is also required to inform the Secretary of State if a mobile installation is converted into a fixed installation, together with a number of other matters relating to changed particulars in relation to the installation.

While these registration requirements are not immediately related to the construction and operational standards of the installation, they clearly inform the enforcement agency of the presence and purported condition of the installation and put the agency on notice that the installation ought to have complied with the construction and survey Regulations and is in any case liable to inspection.

Another 'start up' requirement of the British regulatory system is that if the installation is manned the owner must, in accordance with S4 of the Mineral Workings Act, appoint an installation manager who must always be present on board except under stated conditions. The owner of the installation is required to appoint to be installation manager:

(a) a person who, to the best of the knowledge and belief of the owner, has the skills and competence suitable for the appointment; and

(b) another, or others to act where necessary in place of the installation manager

and shall inform the Secretary of State of any appointment under this sub-section by giving notice in the prescribed form and containing the prescribed details.

The notice of appointment must be given in the form required by the Offshore Installations (Managers) Regulations 1972. The installation must also maintain a logbook as required by the Offshore Installations (Logbooks and Registration of Death) Regulations 1972.

Norwegian legislation

The former system
Historically, Norway, when facing the same problems approached them in a somewhat different manner. The system was one of 'approvals' under which the licensee could not undertake any operations before he had obtained approval from each of the various enforcement agencies. The approval system involved the agencies in much inspection of the licensee's installations, plant and equipment. In the case of mobile installations the Maritime Directorate was the co-ordinating agency to whom the licensee applied. In the case of fixed production platforms the Petroleum Directorate had authority in respect of petroleum laws and delegated authority from the Ministry of Local Government and Labour in respect of the Working Environment Act.

As far as Norwegian-registered mobile installations were concerned the Maritime Directorate operated under what has been described as a 'dual track' system (as is still reflected in the system described in the *Red Book*). Under this system the Norwegian-registered mobile had to satisfy the Maritime Directorate both in respect of petroleum laws and shipping laws made specially for Norwegian mobile installations.

The present system

The role of the Maritime Directorate is now almost entirely confined to approving the structure of Norwegian-registered mobile installations and ensuring these installations operate to the standards of special shipping laws made for them: it has jurisdiction in this respect on a world-wide basis.

Under the Royal Decree of 28 June 1985 the overall responsibility for ensuring the licensee's compliance with the petroleum regime on the Norwegian continental shelf has been delegated to the Norwegian Petroleum Directorate, though this agency may seek the advice of other agencies with specialist expertise.

The present system, which is entirely dependent on *consents*, is a radical departure from the previous system of *approvals*. Under the new system the responsibility for devising a safe system of operation rests squarely on the holder of the licence (which will have been obtained in accordance with the petroleum legislation). He has to satisfy the Petroleum Directorate that he will comply with the legislative requirements while operating to a safe system. The Petroleum Directorate makes its decisions and grants the necessary consents almost entirely on the licensee's documentation.

For purposes of the petroleum regime, Norway no longer specifies that the licensee must involve any external agencies, either governmental or private (such as the classification societies), in certifying either the installation or its plant and equipment. In practice, of course, especially for complying with flag state requirements and for insurance purposes, the installation is almost certain to be subject to a system of inspection and approval by external agencies – indeed, the Maritime Directorate will be performing this function in respect of Norwegian-registered mobile installations.

For the purposes of the petroleum regime the licensee is expected to present a system which includes assurance of compliance with regulatory standards (which for the most part are still those which precede the Act of 22 March 1985). Once he has obtained consents the licensee will be expected to conduct audits to ensure that the regulatory standards are honoured: the Petroleum Directorate will inspect to 'spot check' that the licensee is complying with the standards.

System of consents

The system of consents is governed primarily by the Regulations concerning safety of 28 June 1985 made by Royal Decree under the principal Act of 22 March 1985. The system of consents operates on the philosophy of internal control, in that it places on the licensee the responsibility for identifying what is necessary to obtain consents initially; the regulatory standards on which consent is based must be subsequently maintained through the system of internal control set out in the Regulations concerning internal control which were made by Royal Decree on 28 June 1985.

Consent is required at each phase of offshore operation under the licence. The system is well described in the 'Supplementary Description of the Regulatory Supervision' contained in the 1987 edition of Volume 2 of the *White Book* published by the Petroleum Directorate.

The five phases
The following phases are recognised:

Exploration phase	– shallow drilling
	exploration drilling
Field evaluation	– preliminary
phase	conceptual studies
	further exploration drilling
Engineering phase	– pre-engineering
	detailed engineering
Fabrication phase	– fabrication
	installation
	completion
Operational phase	– start-up
	operation
	rebuilding/alterations
	removal
	relocation

Two Appendices to the description set out the regulatory standards which will be relevant to obtaining consents pursuant to Regulations concerning safety made by Royal Decree on 28 June 1985 in relation to each of the phases.

Appendix 1 *List of regulations relevant to the exploration and field evaluation phases*:
Regulations for mobile drilling platforms issued by the Maritime Directorate 10 September 1973.
Regulations for deck cranes for use on board drilling vessels issued by the Maritime Directorate 31 January 1978.
Regulations for safety arrangements on and below deck on board drilling vessels issued by the Maritime Directorate 31 January 1978.
Regulations for drilling issued by the Petroleum Directorate 23 September 1981.
Regulations concerning qualification requirements for drilling personnel connected with drilling for and production of petroleum issued by the Petroleum Directorate 22 February 1983.
Regulations for the construction and equipping of living quarters on board drilling vessels issued by the Maritime Directorate 11 June 1982.
Temporary Regulations for electrical equipment on installations used for exploration activities issued by the Petroleum Directorate 22 June 1985.
Regulations for life-saving equipment, etc issued by the Maritime Directorate 3 February 1982.
Regulations for approval of survival suits issued by the Maritime Directorate 10 November 1980.
Regulations for the installation and operation of maritime and aeromobile radio

equipment on board drilling vessels and accommodation platforms issued by the Telecommunications Administration 8 September 1980.

Regulations for helicopter decks on board drilling platforms issued by the Civil Aviation Administration 18 April 1973.

Regulations for support vessels used on the Norwegian continental shelf issued by the Ministry of Local Government and Labour 28 December 1983.

Regulations for safety manning in the event of industrial disputes on the Norwegian continental shelf issued by the Ministry of Local Government and Labour 19 March 1982.

Appendix 2 *Regulations and guidelines relevant to petroleum activities*:

Regulations concerning health checks of employees issued by the Ministry of Social Affairs on 1 August 1980.

Regulations for safety manning in the event of industrial disputes on the Norwegian continental shelf issued by the Ministry of Local Government and Labour 19 March 1982.

Regulations for the collection of fees issued by the Ministry of Local Government and Labour 29 June 1984.

Regulations concerning supply vessels used on the Norwegian Continental shelf issued by the Ministry of Local Government and Labour 28 December 1983.

Regulations for the structural design and loadbearing structures issued by the Petroleum Directorate 29 October 1984.

Regulations for cranes on production installations issued by the Petroleum Directorate 25 May 1977.

Regulations for production and auxiliary systems issued by the Petroleum Directorate 3 April 1978.

Regulations for E and P data issued by the Petroleum Directorate 1 August 1978.

Regulations for fixed means of access, etc issued by the Petroleum Directorate 2 April 1979.

Regulations for transfer of personnel issued by the Petroleum Directorate 2 April 1979.

Regulations for drilling issued by the Petroleum Directorate 23 September 1981.

Regulations for qualifications for drilling personnel issued by the Petroleum Directorate 22 February 1983.

Provisional Regulations for diving operations issued by the Petroleum Directorate 1 July 1978.

Provisional Regulations for living quarters on production installations, etc issued by the Petroleum Directorate 2 April 1979.

Guidelines for safety evaluation of platform conceptual design issued by the Petroleum Directorate 1 September 1981.

Guidelines for inspection of primary and secondary structures for production and loading facilities as well as submarine pipe-line systems issued by the Petroleum Directorate 10 November 1982.

Guidelines for specification and operation of dynamically operated diving vessels issued by the Petroleum Directorate 1 May 1983.

Guidelines for area classification issued by the Petroleum Directorate 1 November 1983.

Regulations for life-saving equipment on board permanent installations for production issued by the Maritime Directorate 8 February 1978.

Regulations for approval of survival suits issued by the Maritime Directorate on 10 November 1980.

Regulations for marking of production platforms, etc issued by the Coast Directorate 1 December 1976.

Regulations for the installation and operation of maritime and aeromobile radio equipment on board drilling vessels and accommodation installations issued by the Telecommunications Administration 10 September 1980.

Exploration phase: In this phase consents may be required to carry out exploration activities (in accordance with S8 of the Regulations concerning safety) and exploration drilling (see S9 of the Regulations concerning safety).

Consents for exploration drilling include the use of installations and vessels connected with the activities, as well as relevant operational and contingency-related measures. The detailed regulatory standards which will be relevant are those in Appendix 1.

Field evaluation phase consent to carry out exploration drilling may be required under S5 of the Regulations concerning safety. All the regulatory standards in both Appendices are relevant.

Engineering phase: the engineering phase includes activities connected with procuring technical, economic and administrative information needed to initiate construction and installation work. Consent may be required for a plan of development and operation in accordance with S23 of the Regulations concerning safety and for detailed engineering under S11A.

During the engineering phase the Regulations and guidelines in Appendix 2 apply.

Fabrication phase: this phase includes activities like fabrication, installation, transportation to destination, assembly, and completion of structures and equipment.

During this phase the following main consents must be obtained: consent to fabricate an installation (S11B of the Regulations concerning safety); and consent to install an installation (S11C of the Regulations concerning safety). During this phase the relevant regulations and guidelines are those in Appendix 2.

Operational phase: in addition to normal operations this phase comprises activities initiated in order to ensure that the installation with accompanying systems/components functions in accordance with specifications, and thus is ready for its intended use.

The operational phase further includes activities connected with inspection, maintenance, rebuilding, alteration and improvement as well as eventual removal/repositioning.

For this phase the following consents must be obtained: consent to operate the installation (S11D of the Regulations concerning safety); and consent to rebuild or change intended use of the installation (S11E of the Regulations concerning safety). During the operational phase the Regulations and guidelines in Appendix 2 apply.

This account of the consent system names only the principal consents: in the operational phase in particular the system is broken down into a number of 'partial consents'.

Principles of documentation
The licensee is required to establish and maintain updated documentation files for the various phases. It is part of the licensee's responsibility to establish systems for ensuring that this is done. The documentation must be kept in Norway and be readily available to the Petroleum Directorate.

Before documentation is submitted to the Petroleum Directorate, the licensee is required to verify that the circumstances described in the documentation are in compliance with the regulatory requirements. The documentation must state which persons/departments in the licensee's organisation are responsible for executing, verifying and approving the matters referred to.

The documentation must also set out the internal audits which the licensee has performed or intends within the framework of his internal control system to ensure that the regulatory standards are maintained. Action taken as a result of such audits must be reported (as required by the Royal Decree relating to internal control).

The licensee must also draw up an organisation plan for the phase to which the consent he seeks relates. This plan must be submitted to the Petroleum Directorate together with the application for consent for the phase. The licensee must also have evaluated the activities which are of particular significance for safety: the outcome of these evaluations must be submitted to the Petroleum Directorate.

Where the application involves the use of floating installations or vessels registered as ships (whether Norwegian or foreign), information about the measures which the licensee has taken to evaluate them must also be provided.

References

1   See Chapters 7, 8, 9 and 11 in particular.
2   S1
3   S2: 'they' at this time was the Board of Trade: subsequently the power has been transferred to the Department of Energy.
4   S1(4).
5   This has been done in a series of Continental Shelf (Designation of Areas) Orders which have precisely described the compass points within which the United Kingdom is exercising its rights.
6   S3 *et seq.*
7   S8 *et seq.*
8   Reg 3(2) Schedule 2, Part II.

9   The Norwegians describe this as 'winterising': the extent to which an installation needs to be strengthened to meet weather conditions may well be controversial, since the heavier the installation the more expensive it will be to construct and operate.

10   At the present time he has approved six.

11   Eg Load line certificate.

12   Eg Australian-registered mobile units are subject to merchant shipping laws, and Marine Orders, particularly Part 47, establish detailed requirements relating to the design and operation of mobile units, with special reference to safety.

13   Relevant waters are waters to which the principal Act applies; namely controlled waters (ie tidal and territorial waters) together with designated areas of the continental shelf.

7

Safe Systems in Construction Activities and On Board Installations

Principal legislative provisions

British legislation

Wireless Telegraphy Act 1949.
Mineral Workings (Offshore Installations) Act 1971.
Health and Safety at Work, etc Act 1974.
Oil and Gas (Enterprise) Act 1982.
The Offshore Installations (Construction and Survey) Regulations 1974.
The Offshore Installations (Operational Safety, Health and Welfare Regulations 1976.
The Health and Safety at Work, etc 1974 (Application outside Great Britain) Order 1977.

Norwegian Legislation

Act relating to worker protection and working environment of 4 February 1977.
Regulations supplementing the act on petroleum activities made by Royal Decree of 14 June 1985.
Regulations concerning safety made by Royal Decree of 28 June 1985.
Regulations concerning licensee's internal control made by Royal Decree of 28 June 1985.
Regulations relating to worker protection and working environment, etc made by Royal Decree of 1 June 1979.
Regulations relating to hygiene, medical equipment, etc made by Royal Decree of 25 November 1977.
Provisional standard directive for state-registered nurse stipulated on 1 August 1980.
Standard instruction for physician in charge on fixed installations of May 1982.
Regulations for cranes on production platforms of 25 July 1977.
Regulations on deck cranes, etc for use on board drilling units stipulated on 13 January 1986.
Regulations for transfer of personnel to and from production installations stipulated 2 April 1979.
Provisional regulations for diving stipulated 1 July 1978.
Regulations relating to radio equipment on drilling vessels stipulated 8 September 1980 and 13 January 1986.
Regulations for helicopter decks on drilling platforms of 18 April 1973 and 13 January 1986.
Regulations on control of diving systems of 21 February 1980.

Regulations on working hours of diving personnel on Norwegian ships, drilling platforms, etc of 26 June 1981.
Regulations concerning manning of 23 March 1982.
Regulations concerning qualifications of 23 March 1982.
Regulations concerning the control of diving systems made on 10 April 1984.

The Geneva Convention, and the British and Norwegian legislation to regulate the development of exploration for and exploitation of petroleum resources in the North Sea which followed it, focused on activities conducted on installations for the primary purposes of drilling for and production of petroleum. However, these activities cannot be carried out without considerable support (or servicing) from organisations not directly engaged in the activities of exploration and exploitation of petroleum.

It has already been noted that a mobile installation has two major functions – as a ship and for petroleum exploration – and the performance of these separate functions requires two work teams whose work activities need to be co-ordinated in the interests of achieving a safe system.

Operating a ship does, of course, require 'sailors' with skills over and above those of navigation, not least the ballasting requirements. In a cargo vessel it is common for these additional skills to be exercised by persons who are part of the crew and under the direct command of the ship's captain. In a passenger carrying liner it is not uncommon for services such as catering to be provided by contractors. In this respect the installation has more in common with the passenger liner than the cargo vessel.

In the case of a fixed installation the comparison can in some respects more readily be made between the installation and an onshore industrial establishment, in that it is largely constructed at the place at which it will operate and it will not need to be manned by sailors. Yet the analogy with an onshore installation is inaccurate, both because construction of a fixed platform will only be effected with the services of seagoing craft and because when in operation the platform will need to have much the same degree of self-sufficiency as a ship.

The distinctions between fixed and mobile installations are possibly never more obvious than during the construction phase, as will become apparent in this chapter which will review the systems needed during the construction work, during any structural modifications to the installation and during normal operation. For convenience these matters may be classified as: construction and workover maintenance systems; catering and domestic systems; radio and other communications systems; medical services[1]; and specialist contractual services.

Support systems for emergency and those provided apart from the installation and to maintain links between installation and land (other than radio communications) are considered separately.[2]

Any provision for safe systems of work offshore must include provision for controlling these service activities and also provide for their co-ordination with the primary activities of exploration and exploitation.

Construction and maintenance systems

Both the Norwegian and the British offshore regulatory legislation is focused on 'petroleum activities'. In the British sector the focus is further narrowed by the statutory provision that these activities are carried out 'from, by means of or on an installation'.[3]

This provision therefore creates interesting questions of the point at which a structure, in the course of erection in the North Sea and which is intended to become a fixed installation, can be deemed to have reached the stage in its development where it can be held to have actually become an installation and have thus been brought within the ambit of petroleum-related legislation. The recognition and resolution of these problems place construction activities in a different category from maintenance activities.

Careful maintenance of the installation after its completion and commissioning is clearly important if the operators are to ensure safety and to comply with Regulations, but although maintenance activities may include many of the operations which are found during construction, the legal problems relating to achieving safe systems of maintenance are not substantially different from other situations which involve close co-operation between a head contractor and a sub-contractor. This will be considered below, both in relation to operations on what is acknowledged to be an installation and in relation to the catering and domestic services.

In practical terms it is, however, particularly important that there should be a proper regulatory system of control whenever construction or maintenance work is taking place. It is during these operations when it is most likely there will be the largest number of workers engaged on the installation, and when some of the most dangerous work situations (outside the hydrocarbon-related work) are to be found. The onshore construction industry is notoriously unsafe – offshore there have to be added to the normal hazards of the construction industry the extra hazards of working on or over water in weather that is likely to be inclement in the extreme.

It should, perhaps, be emphasised that construction hazards are largely, but not exclusively, a problem related to the erection of fixed installations. It is not impossible, however, that when a mobile installation is on-site some relatively minor construction work may be carried out on it, but the scale of this work and the problems which it entails are likely to be minimal compared with those associated with the development of fixed platforms.

British legislation

The British legislation of the early 1970s which followed in the wake of the Sea Gem catastrophe[4] arguably conceived installations too narrowly, as structures with the attributes of ships, including a degree of mobility. The Mineral Workings (Offshore Installations) Act and the Regulations made under it, were possibly better suited to the control of structures which, like ships, were both built and assembled in dry dock and on leaving the dock were able to travel out to sea under their own power.

Legislation which presupposed this pattern of operation was not particularly suited to control the construction phase of what was intended to be a fixed platform, where the typical pattern of construction would be for sections of the platform to be built at, or near to a port, then towed out to the site where it is to operate, leaving a hazardous part of the construction phase, including hook-up, to be completed on the continental shelf.[5]

While the Offshore Installations (Operational Safety, Health and Welfare) Regulations 1976 made under the Mineral Workings (Offshore Installations) Act 1971 do make considerable provision both for the control in general of situations where there is more than one organisation engaged in work on an installation and in particular, for the safe execution of construction work, these Regulations are not likely to be in effect during the early stages of offshore construction of a fixed platform. Their real significance is likely to be in the latter stages of construction or if alterations,modifications or repairs are undertaken to an established installation: they apply only if there is an installation 'maintained for the carrying on of any activity to which the Act applies'.[6]

In effect, therefore, the Operational Safety, Health and Welfare Regulations will not apply until persons can be accommodated on the structure under construction, as then such accommodation would be an 'activity' to which the Act applies. However, they will apply from the outset to flotels and crane barges

In effect, therefore, the Operational Safety, Health and Welfare Regulations will not apply until persons can be accommodated on the structure under construction, as then such accommodation would be an 'activity' to which the Act applies. However, they will apply from the outset to flotels and crane barges as installations, because they accommodate persons working on the installation under construction. For the installation where the construction work is being undertaken, only the Health and Safety at Work, etc Act will apply in the initial stages.

Where the Operational Safety, Health and Welfare Regulations do apply, Reg 32 imposes duties on the installation manager, the owner of the installation and the concession owner. Sub-sections (2) and (3) of this Regulation further impose duties which may be especially relevant to sub-contracted employers and their employees:

(2) It shall be the duty of the employer of an employee employed by him for work on or near an offshore installation to ensure that the employee complies with any provision of these Regulations imposing a duty on him or expressly prohibiting him from doing a specified act.

(3) It shall be the duty of every person while on or near an offshore installation –

    (a) not to do anything likely to endanger the safety or health of himself or other persons on or near the installation or to render unsafe any equipment used on or near it;

    (b) to co-operate with his employer, if employed, and any other person on whom a duty or requirement is imposed by these regulations so far as is necessary to enable that duty or requirement to be performed or complied with; and

    (c) to report immediately to the appropriate responsible person or, if no such person be appointed or, if appointed, unavailable, to the installation manager

> any defect in any equipment which appears to him likely to endanger the safety, health or welfare of persons on or near the installation or the safety of the installation and any equipment used with it.

Significantly, Reg 14 also requires that all reasonably practicable steps must be taken to ensure the safety of persons at all places on the installation. This Regulation then specifically refers to the need to ensure, so far as reasonably practicable, the safety of scaffolding, ladders and working platforms. Regulation 34 states that in the event of any of the Regulations being contravened the installation manager, the concession owner and the owner of the installation shall each be guilty of an offence.

The employees of sub-contractors, as all other persons on the installation, are under the control of the installation manager and bound to follow his instructions in matters pertaining to the safety of the installation and of persons on it, in accordance with S5 of the Mineral Workings (Offshore Installations) Act 1971.

That there should be problems related to the application of the Regulations made under the Mineral Working Act to the construction phase of fixed installations is arguably an inevitable consequence of the approach of the British law to observing the restriction on the extension of coastal state jurisdiction which was intended by the Geneva Convention. Any attempt to reconcile control of construction activities on the high seas on the continental shelf with imposed restrictions on jurisdiction, must raise questions as to when such activities are sufficiently related to exploration for and exploitation of petroleum activities for them to be deemed to be part of petroleum activities and thus within the ambit of petroleum laws.

In essence the problem relates to the structure which is intended to become the installation but is not yet even offering accommodation to those engaged in construction. In so far as accommodation vessels are concerned, S24(2)(d) of the Oil and Gas (Enterprise) Act 1982 is a substitution for S1 of the Mineral Workings (Offshore Installations) Act 1971, in order to resolve the problem and bring such 'activity' as the 'provision' of accommodation vessels clearly within the jurisdiction of this Act and the Regulations made under it:

> ... the provision of accommodation for persons who work on or from an installation which is or has been maintained, or is intended to be established, for the carrying on of an activity falling within paragraph (a), (b) or (c) above or this paragraph.

Paragraphs (a), (b) and (c) may be paraphrased as the activities of (a) exploration and exploitation of petroleum resources, (b) the storage of gas and (c) the conveyance of things by pipe.

The actual work of construction will clearly be covered by the Health and Safety at Work, etc Act for the Order in Council of September 1977 (the Health and Safety at Work, etc Act 1974 (Application outside Great Britain) Order), Article 4 extends the provisions of the Act to any installation and any activity on it and activities include (*inter alia*) the following:

> ... construction, reconstruction, alteration, repair, maintenance, cleaning, demolition
> ... and any activity which is immediately preparatory to any of the said activities;

In order to resolve any remaining doubt the Article provides that 'Offshore installation' includes any installation which:

(a) is maintained ... or is intended to be established ...
including any such installation in the course of construction, demolition or dismantling ...

The very comprehensive Offshore Construction Guidance on health and safety applies and indicates standards for compliance with S2 of the Health and Safety at Work, etc Act.

The extension of the provisions of the Health and Safety at Work, etc Act 1974 to installations is most important because they are particularly suited to ensure that safe systems of work exist where a number of organisations are engaged on related work activities. This Act is therefore significant for many of the issues (which will be considered in this and the following chapters) where safety depends on co-ordinating the activities of several organisations engaged in the same, or closely related, tasks.

Section 2 of the Health and Safety at Work, etc Act states:

(1) It shall be the duty of every employer to ensure, so far as is reasonably practicable, the health, safety and welfare at work of all his employees.
(2) Without prejudice to the generality of an employer's duty under the preceding sub-section, the matters to which that duty extends include in particular –
   (a) the provision and maintenance of plant and systems of work that are, so far as is reasonably practicable, safe and without risks to health;
   (b) arrangements for ensuring, so far as reasonably practicable, safety and absence of risks to health in connection with the use, handling, storage and transport of articles and substances;
   (c) the provision of such information, instruction, training and supervision as is necessary to ensure, so far as is reasonably practicable, the health and safety at work of his employees;
   (d) so far as is reasonably practicable as regards any place of work under the employer's control, the maintenance of it in a condition that is safe and without risks to health and the provision and maintenance of means of access to and egress from it that are safe and without such risks;
   (e) the provision and maintenance of a working environment for his employees that is, so far as is reasonably practicable, safe, without risks to health, and adequate as regards facilities and arrangements for their welfare at work.

The provisions of this section impose duties on all employers with employees working on offshore installations. Thus they are equally applicable to the operator of the installation (in relation to his direct employees) and to the employer of the catering staff (see below), and for that matter to any other employers who have their employees on the installation in the course of their employment. Although the provisions of sub-section (2) do not exhaust the general duty in sub-section (1), it is appropriate to consider sub-section (2) carefully because the matters expressly referred to there are particularly significant to the offshore situation.

Section 3(1) of the Act further provides:

It shall be the duty of every employer to conduct his undertaking in such a way as to ensure, so far as is reasonably practicable, that persons not in his employment who may be affected thereby are not thereby exposed to risks to their health or safety.

Section 3, taken in conjunction with S2, in effect imposes duties on each employer (both head contractor and sub-contractors) to ensure, so far as is reasonably practicable, that the activities for which he is responsible do not endanger other persons who are at work. The impact of these provisions in situations where work has been sub-contracted was considered by the Court of Appeal (Criminal Division) in the important case of R v *Swan Hunter Shipbuilders Ltd*.[7]

The criminal prosecution arose out of a catastrophe which was also the subject of an official inquiry.[8] The catastrophe occurred when fire broke out in *HMS Glasgow*, a ship under construction in Swan Hunter's yard. The fire started when a welder, working in a confined space in the bowels of the ship, struck his arc with his welding torch. The atmosphere was oxygen enriched so the fire was very intense and it spread with such rapidity that eight men were entrapped and killed. It emerged that an employee of Telemeter Installations Ltd (a sub-sub-contractor of Swan Hunter) had failed to turn off the oxygen supplied for welding operations when he had left work the previous day. Consequently oxygen had been discharging into the atmosphere from a hose attached to an oxygen cylinder throughout the night, until the time when the fire started.

The hazards of using oxygen in poorly ventilated areas were well known to Swan Hunter's safety officer and he had drawn up a *Blue Book* of instructions for users of fuel and oxygen. These instructions emphasised the dangers and set out a number of safety rules. They were issued by Swan Hunter to their own employees but not to employees of other companies working on their premises.

Swan Hunter and Telemeter were each charged with failing to make the provision required by S2 of the Act for the safety of their own employees, and with failing to discharge the duty under S3 which was owed by them to those who were not in their employment. Both defendants were convicted under these charges but only Swan Hunter appealed. The Court of Appeal confirmed their guilt in relation to both Ss2 and 3.

It was held that in order to discharge their duty as employers to their own employees, under S2(2)(c), it was necessary for Swan Hunter to provide information and instruction to the visiting workers. Section 2(2)(c) required the employer to do what was reasonably practicable to influence the conduct of the visitors to ensure that their conduct did not endanger Swan Hunter's employees. Section 3 required Swan Hunter to give the same information (ie the *Blue Book*) to the visitors to enable the visitors to take care for their own safety.[9]

In addition to illustrating the interlocked responsibilities of employers whose contractual arrangements lead to groups of employees working in close proximity to, and interdependence on, each other, the case established the overriding duty of the head contractor to co-ordinate the operations of the various workforces so as to ensure the safety of all. It also established that the concept of safe systems of work which had for long been fundamental to employers' liability litigation in the civil courts, had, by the Health and Safety at Work, etc Act been made a part of the criminal law.[10]

It is possibly no coincidence that this most important case, concerning the responsibilities of employers to those who are not in their direct employment, arose out of an organisational arrangement of a kind which is extremely common in the construction industry, and it must be as significant to construction and maintenance activities offshore as it is to those which are conducted on land.

The influence of this decision can be seen in Reg 4(2)(f) of the Merchant Shipping (Health and Safety: General Duties) Regulations 1984, which imposes duties upon employers to:

> ... collaborate with others who employ persons who are at any time in the course of their employment aboard a ship or engaged in loading or unloading activities, to protect the health and safety of all persons aboard that ship.

Where regulatory duties, with criminal sanctions, are imposed upon employing organisations in circumstances where the day-to-day control is vested in management, it commonly falls within the scope of the contract of employment of one or more managers to set up and maintain the systems which the law requires the employer to operate. In such cases, difficult legal and factual questions may arise as to where criminal responsibility should rest in the event of the regulatory standards not being observed.

In the offshore context, the Mineral Workings Act places on the installation manager the responsibility for the maintenance of safety on the installation and gives him the authority to require that others obey his command. It is possible therefore that nice questions could fall to be decided in a case of a breach of the general duties under the Health and Safety at Work, etc Act, as to whether liability for this breach should be imposed upon the manager or his employer (for further detail, see Chapter 4).

Norwegian legislation

The framework act of 22 March 1985 is primarily concerned with petroleum activities on the Norwegian continental shelf and the presence of an installation is not a prerequisite of this regulatory control. Indeed, S1 of the Act, having stated that it applies to petroleum activities, continues as a secondary matter: 'The Act *also* applies to installations for such activities ...'[11]

The wider scope of the Norwegian law, as compared with the British Continental Shelf Act, even as now amended by the Oil and Gas (Enterprise) Act, enables the avoidance of some of the difficulties which are inherent in the British system in the early stages of installation construction. The Act of 22 March sets out a framework which covers the whole sequence of activities from the granting of exploration licences through to full production. Building on this, framework Regulations concerning safety made by Royal Decree of 28 June 1985 deal, in S10, with the duty of the licensee to draw up plans for development and operation of petroleum deposits and S11 sets out the main consents which have to be obtained for development and operation. Consents which have to be obtained include consent for fabrication of an installation, consent to install an installation and consent for major rebuilding or changes in operation of the installation.

Section 6 of the Regulations concerning the licensee's internal control made on 28 June 1985, describes the system of presentation of documentation which must be complied with by the licensee during exploration and exploitation. Under this section the licensee is required to provide a description of his organisation for the construction phase of an installation and it is further provided that the licensee is bound by the system he has described.

During the construction phase the working conditions of construction workers are governed by the Act of 4 February, in so far as that has been extended to offshore workers by the Royal Decree of 1 June 1979. This Act (as described below under catering workers) makes provision not only for the working hours and conditions of work which employers must provide for their employees but also places upon contractors the obligation to organise their work to ensure that it interfaces with the work of other employers to produce a safe system.

As has been noted above, the system of consents applies also to rebuilding operations, so under the Norwegian system there is no substantial difference between the regulatory arrangements for construction and those for maintenance. Workers engaged in maintenance short of major rebuilding would still be protected by the Act of 4 February and the obligations which it places on employers to co-operate in the interests of safety.

In spite of the relative simplicity of the Norwegian system as set out here, it is worth noting that it has not escaped the problems which the British system has faced in relation to the control of ships which are servicing offshore operations. Section 1 of the Act of 22 March 1985 expressly states that installations do not comprise supply and stand-by vessels. This was elaborated in a letter to the Energy Committee of the Storting in relation to the Act. The 'petroleum functions' performed by vessels, as has been identified already, are covered by the Act as petroleum activity. Of course, in this context 'activity' must be distinguished from 'construction' – construction of vessels is beyond the present discussion.

There might, nevertheless, be questions about whether the activities of such vessels might be directly controlled. This, it has been noted, is likely to be a particularly important question during the construction phase when the installation itself is not capable of supporting the workforce engaged upon it. The answer to this problem would appear to lie in the control which the Act and Regulations made under it impose upon the licensee.

The licensee will need, in his documentation, to demonstrate the support systems which he is providing for his workforce. The Norwegian authorities, while having no system of direct sanctions against the ships themselves, can withhold consent from a licensee whose arrangements are not deemed to be satisfactory. Specifically, S12 of the Regulations concerning safety of 28 June 1985 empowers the Minister to require the licensee to obtain consent before using service, construction, crane and other vessels. Section 30 of these Regulations also enables the making of more detailed rules concerning ships used in petroleum activities.

It should, however, be further noted that S65 of the Act of 22 March also suggests that 'coercive measures' may be taken against any vessel which violates any directive issued under the Act. Such vessels may be expelled from the shelf or seized and brought to a Norwegian port. This provision might not ensure satisfactory working conditions on foreign flag ships but, if effective, could ensure that they were not employed on the Norwegian shelf. It is hard to conceive of the circumstances in which these measures would be taken, since seizure of a foreign flag ship would entail a nice balance of Norway's shelf rights against the general concept of freedom of the high seas which is protected by the Geneva Convention, under which the shelf rights of coastal states are exercised.

Catering and domestic services

It is normal practice for any shore-based industrial activity to sub-contract catering, cleaning, and often maintenance, activities to other organisations. The only ways in which the offshore industry differs from the onshore production site are: first, the circumstance that the workforce is living on the installation magnifies the size of the task; and second, the hazards created by both the weather and the petroleum activity itself call for a more careful consideration of the safety of the service workers than is normally required when dealing with, for example, the caterers employed under sub-contract to work in a factory canteen. Moreover the service workers, like the petroleum workers themselves, will be employed offshore for considerable periods of time and provision for their working conditions and environment must take this into account.

British legislation

In the British sector of the North Sea continental shelf, the Offshore Installations (Operational Safety, Health and Welfare) Regulations 1976 Part III is concerned with the health of persons working offshore. Regulation 25 makes provision for an adequate supply of clean, wholesome drinking water on installations and Reg 26 provides:

> All provisions for consumption by persons on an offshore installation shall be fit for human consumption, palatable and of good quality.

The interrelated duties imposed by these Regulations for the purposes of achieving safe systems of work (discussed in relation to construction) are equally relevant in the present context. Indeed they could be invoked to protect catering and domestic staff from hazards created during construction, while at the same time ensuring that wholesome food and drinking water be provided to construction workers.

Similarly, the provisions of the Health and Safety at Work, etc Act are equally relevant. The case of Swan Hunter (set out when discussing construction above) concerned the employer's duty to communicate information to persons who were not in his own employment in order to discharge the duty which S2(2)(c) imposed upon him for the safety of his own employees. It would appear that the same principles could apply in relation to other of the requirements of S2(2). It would seem to be relevant in the present context to suggest, for example, that the head contractor (most probably the installation owner) would be required to ensure: that cooking and refrigeration equipment used by the caterers was neither unsafe nor a danger to health, or he might otherwise be in breach of the duty which he owed to his own employees under S2(2)(a); and that the cleaning staff left floors and stairways in a safe condition, otherwise he might be in breach of S2(2)(d). It is arguable that if the food provided by the contracted caterers proved unappetising, the employer might be in breach of S2(2)(e) which relates to 'facilities and arrangements for their welfare at work'.

On an installation, the owner would also have these responsibilities by virtue of the Operational Safety, Health and Welfare Regulations. Indirectly, the certification authorities would also bear responsibility under the Construction and Survey Regulations.

Moreover, the facts on which liability was imposed under Swan Hunter can leave little doubt that – quite apart from any particular requirements contained under any of the Offshore Installations Regulations – the owner would be in breach of the general duties under the Health and Safety at Work, etc Act if he failed to instruct and inform sub-contractors, employees and other visiting workers, of the safety procedures and emergency systems which he has adopted for the safety of all who are on board the installation. Moreover, the Operational Safety, Health and Welfare Regulations concerning provisions would not apply to food prepared on an installation in the initial stages of construction, although they would apply to food provided on a flotel for the construction workers. At this initial stage, therefore, the Health and Safety at Work, etc Act's general duties would be especially important.

Substance is given to the importance of this proposition by the decision recently given by the Court of Appeal in *R v Mara*,[12] where a cleaning contractor working onshore was found to be in breach of S3(1): an employee at the store he was contracted to clean was electrocuted while using a machine left at the store by the contractor.[13]

Norwegian legislation

Since the introduction of the present regulatory system based on the Act of 22 March 1985 pertaining to petroleum activities, the control of servicing activities associated with petroleum activities on the Norwegian continental shelf has been – like the primary petroleum activities themselves – in effect governed by this legislation and enforced by the Petroleum Directorate; although for the immediate future the detailed regulatory standards are still those made under the former regulatory system. The legislation, like that which it largely supersedes, places primary responsibility for ensuring safe systems offshore upon the licensee. Under the present regime, however, responsibility rests with the licensee both to set up the systems necessary to comply with the regulatory standards and thereafter to ensure, through a system of internal audits, that the necessary standards are observed. Thus, S49 of the principal Act requires that the licensee:

> ... shall ensure that all persons engaged in the activities possess the necessary qualifications to perform in a prudent manner the work they have been assigned. Training shall be given to the extent necessary.

This provision is not qualified by limiting the duty to those who are in the direct employment of the licensee. In the Norwegian, as in the British sector, catering and other service and maintenance tasks are likely to be sub-contracted, but in that sector, as in the British, responsibility for ensuring that this servicing task is properly performed rests with a principal, or head, contractor rather than with the immediate employer of those engaged in the service activities concerned. As has been noted elsewhere, however, in this instance as in most matters in the Norwegian sector, responsibility rests with the licensee, whether or not the licensee is the actual operator.

Also, in the Norwegian system the liability is under the licence and therefore it is a liability more closely related to civil liability for breach of contract than the criminal liability which is directly associated with breach of duty in the British regulatory system. Nevertheless, in the Norwegian system breach of a condition or Regulation under a licence may incur criminal liability by virtue of Ss56, 57 and 66 of the Act of 22 March 1985.

The current emphasis on the licensee's system of internal control – as provided for by the Royal Decree of 28 June 1985 – is sufficient to place on the licensee the duty to satisfy the Petroleum Directorate that the installation is adequately staffed to provide the catering and other services which the overall manning requires.

The licensee's responsibility extends to ensuring that the installation is equipped and maintained to the hygiene standards for the storage and preparation of food which are set out by the Health Directorate in the Royal Decree of 25 November 1977. The licensee therefore needs to ensure that sub-contractors comply with the statutory standards if he is not to endanger his licence. In this case, however, as tends to be the pattern in the Norwegian regulatory system, provision is made for the punishment of any actual wrongdoer. Section 27 stipulates:

> Wilful or negligent violation of the provisions of this Decree or Regulations issued pursuant thereto shall be subject to fines under S339 No 2 of the Norwegian Penal Code, except where more severe penal sanctions are applicable.

Since the Act of 1985 prosecution would most likely be taken directly under that Act.

Similarly, the detailed provisions concerning the equipment of living quarters made for production installations on 2 April 1979, and for drilling and other mobile installations on 11 June 1982, place duties upon the licensee to ensure the compliance of others. Section 2.4.1 provides:

> The licensee is responsible for ensuring that any person who performs work for him, whether personally or through employees contractors, or sub-contractors, observes the provisions of these regulations and or orders pursuant thereto.

Once again the responsibility which the licensee now has to conduct audits as part of the system of internal control, effectively supersedes this direct and narrows reference to his responsibility.

The safety and welfare of sub-contractors' labour rests with the licensee also:

he must clearly take them into account when considering the accommodation, provisioning and life-saving provisions on the installation.

The Norwegian regulatory system is similar to that in the British sector in that in both instances the general onshore regulatory provision for safe systems of work applies offshore. The Norwegian Act of 4 February 1977 places, as does the British Health and Safety at Work, etc Act, primary responsibility for the systems which prevail at the workplace upon the employer (S14). Section 15 provides expressly for the situation where there is more than one employer at a particular workplace:

> When two or more employers are conducting activities simultaneously at one and the same workplace:
>
> a) each employer shall ensure that his own activities and his employees' work are arranged and carried out so that the other employers' employees are also protected in accordance with the rules of this Act.
>
> b) each employer shall co-operate to provide a fully satisfactory working environment for all employees at the workplace;
>
> c) the principal enterprise shall be responsible for co-ordinating the safety and environment at work of each enterprise.

It would appear that the intention is very similar to that of Ss2 and 3 of the Health and Safety at Work, etc Act and the interpretation placed upon these sections in the Swan Hunter case, even though the Norwegian legislation makes no direct reference to the need for the head contractor to instruct the employees of the sub-contractor.[14] Again, the system of internal control effectively covers the situation offshore.

Norwegian mobile installations which are registered as Norwegian flag ships are subject to shipping laws and in particular to two sets of Regulations laid down on 23 March 1982, concerning manning and qualification requirements for personnel on mobile drilling units. The manning Regulations specifically state that the catering section leader is subordinate to the platform manager. It is, however, the catering section leader's responsibility to ensure that all personnel under his supervision possess the necessary qualifications and experience properly to carry out the work assigned to them, from both a professional and a safety point of view. He is also responsible for the general safety training of the personnel within his area of responsibility. Section 8 of the qualification Regulations states the educational and training requirements for each grade of catering staff.

Radio and other communications systems

A very important support system provided on board vessels and installations is the radio link between the vessel/installation and other vessels/installations and between the offshore base and its shore base. By the time that offshore development of the petroleum industry in the North Sea was undertaken in the 1960s, the immense value of radio communications systems to the shipping industry was fully appreciated. The necessity of maintaining close installation-to-shore links and links between installations and support systems, such as stand-by vessels, was so clearly an operational necessity that the special need to maintain such links in the interests of safety need hardly have been provided for in legislation. The task of operating such equipment is so central to the offshore operation that it is normally undertaken by persons in the direct employment of the owner. In this area commercial and safety interests would not appear to be in any way in conflict. Nevertheless, both the British and the Norwegian regulatory systems do make express references to telegraphic communications.

British legislation

The Wireless Telegraphy Act 1949 is primarily intended to control the unauthorised establishment of wireless telegraphy stations. It makes it an offence for any person to install or use any apparatus for wireless telegraphy except under and in accordance with a licence[15] and enables the making of Regulations to control the use of any radio station,[16] stipulating the type of equipment which may be used.[17] The Act is therefore primarily directed to the control of pirate radio systems. However, it also enables the establishment of a system for examining the competence of persons wishing to fill the post of radio operator.[18]

Originally, the Act applied only to radio stations operated in the UK, its territorial waters and British-registered ships and aircraft.[19] Section 6 of the Continental Shelf Act 1964 extended the territorial jurisdiction of the Wireless Telegraphy Act by providing:

An Order in Council ... may make provision for treating for the purposes of the Wireless Telegraphy Act 1949 and any Regulations made thereunder any installation in an area or part with respect to which provision is made under that section and any waters within five hundred metres of such an installation as if they were situated in such part of the United Kingdom as may be specified in the Order.[20]

As soon as the installation is subject to the Offshore Installations (Operational Safety, Health and Welfare) Regulations 1976, Reg 18 requires that any installation which is normally manned shall be properly equipped with signalling equipment:

... there shall be provided on every offshore installation such signalling equipment as will enable effective communication by radiotelephone, on appropriate channels, to be maintained between the installation and radio stations in the United Kingdom and between it and vessels, helicopters and other offshore installations.

Equipment thus provided would, by virtue of the extension offshore of the Wireless Telegraphy Act be subject to the standards laid down under that Act. The extension offshore of this Act is also relevant in connection with Reg 19 of the Operational Safety, Health and Welfare Regulations, concerning radiotelephone operators. This Regulation requires:

There shall be present on every offshore installation at any time when it is manned at least one person fully trained to be the radiotelephone operator who is the holder of a certificate of competence valid with respect to the equipment provided on the installation under Regulation (1) above issued by the Secretary of State under S7(1) of the Wireless Telegraphy Act 1949.

Safety Training Guidance Note No 1, issued by the Department of Energy indicates which certificates are acceptable.

The availability of radio communications systems and the regulatory requirement that they be installed on manned installations must be relevant to the interpretation of other regulatory provisions such as that contained in Reg 9 of the Offshore Installations (Inspectors and Casualties) Regulations 1973, which requires that when a casualty has occurred the manager of the installation on or near to which it occurred: '... shall, in the most expeditious manner practicable, immediately inform the owner of the installation of its occurrence ...'

The importance of good communications systems is evident when setting out systems for dealing with emergencies and disasters. It is therefore provided also in Reg 18 of the Safety, Health and Welfare Regulations that no building or room in which any relevant radio equipment is installed shall be situated in a 'hazardous area' (ie an area where there is likely to be danger of fire or explosion).[21] Regulation 18 also provides that there shall be an instruction card giving a clear summary of the radiotelephone distress, emergency and safety procedures displayed in full view of the radiotelephone operating position.[22]

Regulation 18(5) requires that there has to be provided signalling equipment when normally unmanned installations have persons on board.[23]

The above Regulations are again only applicable when the installation is sufficiently operational for the mineral workings legislation to be invoked. Until this stage only the Health and Safety at Work, etc Act is applicable. Radio communication between installations and stand-by vessels is governed by Reg 10(2) of the Offshore Installations (Emergency Procedures) Regulations. It ought also to be noted that there is no regulatory provision in these matters for pipe-laying barges.

Norwegian legislation

The regulatory system introduced under the Act of 22 March 1985 pertaining to petroleum activities and the Royal Decrees made under it, place far more emphasis on the licensee's responsibility to produce safe systems of work through a system of internal control by the licensee of his own organisation, than was previously the case in the Norwegian system or, indeed, arguably is the case in the British system.

Regulations concerning safety made by Royal Decree of 28 June 1985 to implement the principal Act of 22 March, lay down a broad requirement concerning telecommunications. Section 40 of the Regulations states:

> The installation shall be equipped with telecommunications for necessary communication, in order that installation and operation may take place in a safe manner.

> The Ministry may require that telecommunication systems for the purpose of remote control are installed.

This would appear to be the only express requirement as to the establishment of a system for communication within the current regulatory framework. However, the earlier Regulations relating to the installation and use of maritime and aeromobile radio equipment on board drilling vessels which were made on 8 September 1980, remain in force in so far as they were not expressly annulled by the Royal Decree of 28 June 1985.[24] These Regulations lay down a fairly detailed system for the inspection of the radio installation; they also regulate for the location of radio stations, their operation and the equipment which is mandatory. Finally, they make special provision for fire extinguishing equipment in and in the vicinity of the radio room.

It must also be borne in mind that under S11 of the new safety Regulations[25] the licensee has to obtain certain consents in order to develop and operate, and it would appear unlikely that the relevant consents would be granted unless the licensee's documentation presented under these Regulations was satisfactory. Moreover, under S7 of these Regulations it is a requirement that the licensee's system of control shall ensure that the licensee's and contractor's employees are given necessary training.

In the official explanation of the new supervisory system stipulated under the Royal Decree of 28 June 1985, the Norwegian Telecommunications Administration is specifically named as an agency which will give the Petroleum Directorate supervisory assistance:

> Assistance in the areas not regulated in or in pursuance of, the telecommunications legislation. A more precise definition of the areas/aspects on which assistance will be given, will be outlined in the agreement.

Norwegian mobile installations, registered under the Norwegian flag, are also subject to Regulations issued by the Norwegian Maritime Directorate on 13 January 1986, concerning the installation and use of maritime and aeromobile radio equipment. Under these Regulations the licensee requires a special licence in respect of the radio communication activity. These Regulations give considerable detail about the situation of the radio station on the installations and as to how it should be operated. Section 4 of the Regulations concerning qualification for personnel on Norwegian drilling units and other mobile offshore installations issued on 23 March 1982, require that the radio officer shall have a valid Radio Telegraphy Certificate issued by the Norwegian Telecommunications Administration.

As in the British system, of course, the major significance of the telecommunications system is in respect of contingency plans for catastrophe.

References

1   Dealt with under personnel in Chapter 9 and also under emergencies in Chapter 13.
2   Provision for emergencies is considered in Chapter 13; other servicing and support systems in Chapter 8.
3   S1 of the Mineral Workings (Offshore Installations) Act 1971 as amended by S24(1) of the Oil and Gas (Enterprise) Act 1982.
4   Report of the *Inquiry into the Causes of the Accident to the Drilling Rig Sea Gem* HMSO, Cmnd 3409, 1969.

5 Construction may then be carried out from a barge. This will be covered by the Health and Safety at Work, etc Act by reason of the Order in Council of 1977. See below for discussion of this Act.

6 See application clause of the Application of Statutory Instruments Regulations 1984.

7 [1981] ICR 831; see also *R v Mara* [1987] 1 WLR 87.

8 *Report on fire on HMS Glasgow, 23 September 1976* HMSO, 1979.

9 See Brenda Barrett, The Need to Communicate about Safety Precautions, 45 *Modern Law Review 338* 1982.

10 See in particular *McArdle v Andmac Roofing Co and Others* [1967] 1 WLR especially Edmund Davies LJ at pp367–369.

11 Authors' italics.

12 See 8.

13 Once again the owner would have responsibilities under mineral workings legislation.

14 But see S49 of the principal Act above which as far as petroleum activities are concerned places a clear obligation to train on the licensee, apparently irrespective of where the contract of employment lies.

15 S1.

16 S3.

17 S10.

18 S7.

19 S6.

20 Orders in Council are now made in accordance with S23 of the Oil and Gas (Enterprise) Act 1982.

21 Reg 1 of the Offshore Installations (Operational Safety Health and Welfare) Regulations.

22 The importance of radio communication in systems for dealing with emergencies is dealt with more fully in Chapter 13.

23 See following chapter and Chapter 13 for communications between installations and stand-bys in emergencies and also between installations and helicopters.

24 Annuls Ss3 and 4.

25 By Royal Decree of 28 June 1985.

8

Safe Systems: Relationships between Installations and Support Systems

Principal legislative provisions

British legislation

Wireless Telegraphy Act 1949.
Mineral Workings (Offshore Installations) Act 1971.
Health and Safety at Work, etc Act 1974.
Petroleum and Submarine Pipe-lines Act 1975.
Oil and Gas (Enterprise) Act 1982.
The Offshore Installations (Logbooks and Registration of Death) Regulations, 1972.
The Offshore Installations (Construction and Survey) Regulations 1974.
The Offshore Installations (Operational Safety, Health and Welfare) Regulations 1976.
The Offshore Installations (Emergency Procedures) Regulations 1976.
The Health and Safety at Work, etc Act 1974 (Application outside Great Britain) Order 1977.
Diving Operations at Work Regulations 1981.
Submarine Pipe-lines Safety Regulations 1982.

Norwegian legislation

Royal Decree concerning control of foreigners of 28 August 1967.
Act relating to worker protection and working environment of 4 February 1977.
Regulations supplementing the act on petroleum activities made by Royal Decree of 14th June 1985.
Regulations concerning safety made by Royal Decree of 28 June 1985.
Regulations concerning licensee's internal control made by Royal Decree of 28 June 1985.
Regulations for safety delegates and working environment committees made by Royal Decree of 29 April 1977.
Regulations relating to worker protection and working environment, etc made by Royal Decree of 13 September 1985.
Regulations relating to hygiene, medical equipment, etc made by Royal Decree of 25 November 1977.
Provisional standard directive for state-registered nurse of 1 August 1980.
Standard instruction for physician in charge on fixed installations of May 1982.
Regulations for cranes on production platforms, stipulated 25 July, 1977.

Regulations on deck cranes, etc for use on board drilling units stipulated 31 January 1978 and 13 January 1986.

Regulations for transfer of personnel to and from production installations stipulated 2 April 1979.

Provisional Regulations for diving stipulated 1 July 1978.

Regulations relating to radio equipment on drilling vessels of 8 September 1980 and 13 January 1986.

Regulations for helicopter decks on drilling platforms of 18 April 1973 and 13 January 1986.

Regulations on working hours on Norwegian drilling platforms made 19 August 1977.

Regulations on the keeping of a journal of working hours on drilling units of 11 May 1978.

Regulations on control of diving systems of 21 February 1980.

Regulations on working hours of diving personnel on Norwegian ships, drilling platforms, etc of 26 June 1981.

Regulations concerning the control of diving systems made on 10 April 1984.

Regulations concerning protection supervisors and environment committees of 15 November 1976.

Regulations concerning mustering of employees of 16 June 1975.

In the previous chapter, support systems during the construction stage and on board the installation were considered. It is also necessary for the installation to have good links with, and receive support from the shore base of its own organisation. The absolute, or relative immobility of the installation make this a more acute requirement than for a ship. Ships have traditionally expected to be self-supporting, with minimal contacts with shore between calls at ports to *inter alia* take on supplies.

Both the Norwegian and the British regulatory systems for petroleum activities have had to cater for the reality that installations are in practice in more immediate and constant dependence on links with the shore than is the case for ships. This greater reliance on links with land may reflect differences between the shipping and the petroleum functions, but may also reflect the fact that offshore petroleum activities have only come into existence in an age when there are technical possibilities for rapid and efficient communication.

The support systems needed for emergencies, and those provided apart from the installation, and to maintain links between installation and land base, are considered separately.[1] The subject matter of the present chapter is: 1) transport systems for the conveyance of goods and persons between shore and installation by: a) helicopter; b) ship;[2] 2) pipe-line activities; 3) diving activities; 4) shore bases.[3]

Some of the issues involved have already been considered in Chapter 3 and others will be considered in Chapter 10. In Chapter 3 the legal provisions made for the control of the interface between shipping and petroleum activities were discussed. Since this book is primarily concerned with petroleum activities, the regulation of the operation of ships will not be further discussed here.

This chapter will be concerned only with the ways in which service activities are brought within petroleum laws: either because those laws impose obligations directly upon those engaged in servicing; or because they impose on those who operate installations, obligations with which, in practice, they can only comply by entering into contractual arrangements with other organisations. A good example of the latter situation is in relation to stand-by vessels: British petroleum law places no obligations on those who own or operate a vessel operating as a stand-by vessel to an installation in the British sector, but the Offshore Installations (Emergency Procedures) Regulations 1976 Reg 10 requires:

(1) There shall be present within five nautical miles of every offshore installation when it is manned a vessel (in these Regulations referred to as a 'stand-by vessel') ready to give assistance in the event of an emergency on or near the installation and –

    (a) which is capable of accommodating safely on board all persons who may be on the installation at any time; and

    (b) which is equipped to provide first-aid treatment for all such persons.

Regulation 11 places the obligation to ensure that these provisions are complied with on the owner of the installation and the concession owner. The implication is, therefore, that if the operators of the installation do not have the direct authority to control and manage the stand-by vessel they must, as part of their hiring arrangements, stipulate that the vessel meets the standards which the law imposes on the installation.

Transport systems

The significance of ships in carrying persons and goods to and from installations was mentioned earlier, and it may be noted that British merchant shipping law has addressed the problem of the interface between shipping and petroleum laws in transportation of goods between ships and installations.[4] Helicopters also play a significant part in the transportation of people, light goods and emergency supplies.

At the outset it should be remembered that it is the responsibility of the installation manager to maintain a record of the persons on or working from the installation. This record is in practice maintained by entry of the names of persons arriving at or leaving the installation, at the point of arrival on or departure from the installation. The maintenance of this record is an important aspect of any personnel transportation system and incidentally establishes the responsibility of the manager, the licensee (in the Norwegian system) and the owner (in the British system) for the presence on and safety of all persons working on the installation whether they are in the direct employment of the owner, or are working on the installation under sub-contract.

The exact nature of the responsibility is spelt out in the Offshore Installations (Logbooks and Registration of Death) Regulations 1972, Reg 7 of which stipulates:

... there shall also be maintained on the installation a separate continuous record of the persons on or working from the installation which shall include –
(a) the full names of every such person;

(b) the date and time of his arrival and, if he is no longer on or working from the installation, of his departure;

(c) the reason for his presence there;

(d) the name and address of his employer: ...

The Regulation further states that the owner of the installation shall require the manager to notify him at least once every 24 hours of the persons then on or working from the installation, and that the owner shall maintain a duplicate record at a place ashore in the United Kingdom. The shore record must also show the nationality, date of birth and usual residence of the persons concerned and give particulars of their next of kin.

The movement of personnel to and from installations is likely to be largely in two sorts of circumstances: the conveyance to and from the installation of the regular workforce at the beginning and end of each period offshore and the conveyance of specialist workmen who are visiting the installation for the performance of some particular task, such as the installation, maintenance or repair of specialised equipment.

In addition in the British legislation, it is expressly required that in the case of mobile installations the number of persons on board shall be kept to the minimum during movement of the installation.[5] The matter is covered by Reg 8 of the Offshore Installations (Operational Safety, Health and Welfare) Regulations, which provides:

> At a time when an offshore installation is in course of being raised or lowered or dismantled no person who is not essential to the operation shall be thereon and no person who is thereon shall be thereon without the written consent of the installation manager.

There does not appear to be any express stipulation comparable with this in the Norwegian system but safe operation would suggest that this ought to be a matter for the licensee's internal control.

Supply ships

It has already been noted in Chapter 3 that petroleum laws have sought to impose control over ships in the vicinity of installations, particularly in relation to the activities of loading and unloading.

British legislation

The British law provides in S5 of the Mineral Workings (Offshore Installations) Act 1971:

> The manager of an offshore installation shall not permit the installation to be used in any manner, or permit any operation to be carried out on or from the installation, if the seaworthiness or stability of the installation is likely to be endangered by its use in that manner, or by the carrying out of that operation or by its being carried out in the manner proposed ...

The Act further provides that if any person is acting in such a way as to endanger the installation or persons on it, the installation manager is empowered to restrain them. These statutory powers could only be exercised, it would appear, in relation to those on board the installation and thus would entitle the manager in the last instance to restrain those conducting the operation from the installation but, not, it would seem, to give him access to and power over those on board the vessel.

An important part is played by cranes in loading and unloading operations, and there are particular dangers in operating cranes in the heavy seas and poor weather conditions found offshore. When the installation is sufficiently established to be within the Offshore Installations (Operational Safety, Health and Welfare) Regulations 1976, Reg 13 is applicable. This provides:

(1) Every lifting appliance or piece of lifting gear used as or forming part of the equipment of an offshore installation shall be plainly marked with its safe working load or loads as shown on the latest record of thorough examination required under Regulation 6(3) and no lifting appliance or piece of lifting gear shall be used by any person for any load exceeding the safe working load marked thereon.

Sub-section (2) requires that in the case of a multiple sling, the safe working load at different angles of the legs must be marked in the manner described in Schedule 3 of the Regulations.

The Department of Energy's newly issued Training Guidance Note No 5 deals comprehensively with the selection, training and duties of offshore crane drivers.

Regulation 7 of the Operational Safety, Health and Welfare Regulations requires that there shall be provided by the owner of the installation, written instructions specifying practices to be observed to ensure the safety of the installation and the safe use of equipment thereon. Schedule 2 of the Regulations specifies that such instructions shall be provided for the operation of lifting appliances, lifting gear, including use of slings, chains, wire ropes and other lifting tackle. It is the duty of the installation manager to ensure that the activities of any person engaged in any operation or work on, from, or in connection with an installation, are carried out in accordance with the written instructions provided in respect of the installation, and that the relevant part of the instructions is brought to the attention of every such person.

In the initial stage of construction of what is intended to be an installation, the sole legal regime may be the Health and Safety at Work, etc Act 1974. However, the safe operation of cranes is also covered in Offshore Installations (Guidance Notes) Part IV and these are applicable in the construction stage, before the installation is registered.

Norwegian legislation

In the Norwegian system, S43 of the Regulations concerning safety made by Royal Decree on 28 June 1985 places upon the licensee responsibility for providing a safe system for the transportation of persons to and from installations. This responsibility exists whether the method of transportation is ship or helicopter.

Particular provision is made for the transfer of personnel to and from production installations by Regulations of 2 April 1979. These Regulations contain the general provision that any transfer of personnel to or from an installation with no connection by bridges shall normally be performed by the use of helicopter or other approved method of transfer.

To this general provision, however, the exception is allowed that in emergency transfer may be by personnel basket. Such means of transfer may only be used when transfer cannot be delayed due to considerations related to the safety of the personnel or the installation. The circumstances in which this method of transfer may be used are thus so restricted as to make it seem highly unlikely that an emergency would occur when the appropriate conditions prevailed. For example, the basket may not be used when the wind force is above 30 knots, in hours of darkness or periods of poor visibility.

Norwegian petroleum laws have also paid special attention to the particular dangers of using cranes on installations for the movement of goods to and from installations. Two sets of Regulations have been made. Regulations for cranes on production installations made on 25 July 1977 make detailed provisions for the design, use and maintenance of cranes, and require that crane operators have the approved crane operator's licence. Regulations on deck cranes for use on board drilling units make similar and equally detailed provision for both Norwegian and foreign drilling units.[6] Norwegian Regulations for mobile installations also impose responsibilities on the platform manager to ensure that there is adequate supervision of all loading and unloading of ships.[7]

Stand-by vessels

As was noted in Chapter 3, both British and Norwegian law require that every manned installation shall be attended by a stand-by vessel capable of rendering assistance in an emergency. The provisions for stand-by vessels are contained in Regulations concerning emergency procedures; further consideration of their role can therefore appropriately be left to that context.

Helicopters

The employment of helicopters in petroleum activities involves very similar problems to those which arise in connection with the employment of ships: namely, helicopters are subject to a regulatory regime under civil aviation legislation – a system which is quite distinct from and independent of the legislative control of petroleum activities. As was noted in Chapter 3, the interface between two regulatory systems, as will occur when the ship or the helicopter enters the safety zone surrounding the installation, raises questions concerning command structures, with problems as to where ultimate authority rests should conflict arise between the two systems of operation.

The use of helicopters has expanded to the point where they are virtually the sole means of transportation for the offshore workforce between the shore and installations: indeed, Aberdeen (Dyce) Airport is now regarded as the busiest heliport in the world.

There have been a number of tragedies involving helicopters in the British and Norwegian sectors of the continental shelf, and other incidents world-wide involving the movement of oil workers and contractors' personnel. They are all dwarfed by the disaster which overtook a Boeing 234 twin-rotor helicopter, known as the Chinook, on 6 November 1986, while returning from the Brent oil field with 47 persons on board. It was some two miles short of the heliport at Sunburgh, in the Shetland Islands when, without warning, it dropped from 500 feet and crashed into the sea.

The wreckage was raised from the sea-bed and meticulously examined. An interim report of the British Department of Transport's accident investigation team[8] revealed that a catastrophic failure had occurred in part of the forward rotor gear box. This would have led to the two sets of rotor blades becoming desynchronised, with the rear blades colliding with those of the forward rotor.

Civil aviation laws lay down the standards for the construction and maintenance of aircraft, including helicopters, the working conditions and qualifications for those who fly in them, and stipulate the manner in which the aircraft shall be operated. These matters, like the comparable merchant shipping laws, are beyond the scope of this book. The matters to be considered here relate solely to the interface between the rules for the control of the helicopter and those for the control of the installation when the helicopter is in the vicinity of the installation.

British legislation

The Offshore Installations (Construction and Survey) Regulations recognise the importance of helicopters to the support of installations and provide in some detail for the provision of a helicopter landing area on the installation. Schedule 2 Part V provides:

Every helicopter landing area forming part of an offshore installation shall –
(a) be located in a position readily accessible to and from the living accommodation of the installation or any other area of the installation likely to be regularly manned;
(b) be large enough, and have sufficient clear approach and departure paths, to enable any helicopter intended to use the landing area safely to land thereon and to take off therefrom in any wind and weather conditions permitting helicopter operation;
(c) be strong enough to withstand any landing by any helicopter intended to be used;
(d) be provided with a non-slip surface for landing, constructed so that rainwater and fuel spills shall not collect thereon or fall therefrom on to other parts of the installation;
(e) be provided with suitable tie-down points;
(f) be provided with markings and lighting sufficient to enable easy identification of the landing area by day or night and have any obstructions thereon clearly marked and illuminated; and
(g) be equipped with suitable safety nets along the sides thereof over which persons might fall.

The Offshore Installations (Operational Safety, Health and Welfare) Regulations set out the responsibilities of those engaged in petroleum activities in respect of visits of helicopters to, their landing on, and departure from an installation. The relevant Regulations (Regs 21–23) contain a considerable amount of detail concerning the following.

Helicopter landing officer: every manned installation is required to have on board a competent person appointed by the installation manager to be responsible for the control of helicopter operations and all persons engaged in helicopter operations or near to the landing area are to be under the control of the helicopter landing officer.

Helicopter operations: Reg 22 stresses the importance of radio communications between the helicopter and the installation – indeed no helicopter must become airborne with the objective of landing on an installation unless radio communication has been established between the helicopter and the installation. Broadly, the intention is that radio communication will be maintained between helicopter and installation throughout landing and take-off. It is the duty of the installation manager to provide to the helicopter information about the weather at the installation.

The installation must be equipped to receive the helicopter, including specified equipment, such as chocks and tie-down ropes, which are required in normal circumstances and equipment needed for use in the event of an accident, including a large axe and heavy-duty hacksaw.

Any installation which is more than 50 nautical miles from land must have sufficient fuel to enable a helicopter to fly from the installation to land and safe and efficient equipment for refuelling the helicopter.

It is the duty of the landing officer to notify the installation manager of any shortage of or need to replace the equipment which the installation is required to carry.

It is also the landing officer's duty to ensure that before a helicopter lands on or takes off from an installation the landing area is clear of obstruction (including snow, heavy spray or sea), that cranes have ceased to operate, that no person whose presence is not necessary to the helicopter operations is in the vicinity, that the fire-fighting equipment for the area is manned, the stand-by is informed that helicopter operations are to take place and that safety nets are secured.

In fact the requirement that radio communication will be maintained between helicopter and installation throughout take-off and landing is likely to be impracticable. The Department of Energy's Training Guidance Note No 2 gives guidance on the selection, training and appointment of persons as helicopter landing officers.

The control of helicopter landing operations must be placed in the context of S5 of the Mineral Workings (Offshore Installations) Act, which gives the installation manager overall authority over all persons and operations on or about the installation. Section 5(4), in particular, enables the manager to refuse permission for any operation to be carried out if the seaworthiness or stability of the installation might be endangered.

Section 43 of the Regulations concerning safety which were made by Royal Decree on 28 June 1985 requires, in general terms, that the licensee shall provide transportation facilities and equipment for transfer of personnel, including landing areas. It is now clear that in future, setting up safe systems for the landing of helicopters will be among the matters to which the licensee will be expected to have regard when laying out his system of internal control under the Regulations on internal control made by Royal Decree on 28 June 1985. However, the Norwegian Civil Aviation Authority is one of the agencies with whom the Petroleum Directorate has an agreement for assistance when the Petroleum Directorate is selecting and enforcing standards for installations. Standards are contained in Regulations for helicopter decks on drilling platforms. These Regulations contain substantially the same material on construction and operation of helicopter landing areas as has been noted in the British legislation, including the appointment of a landing officer.

Section 47 of the principal Act of 22 March 1965 stipulates that no unauthorised helicopters may be present in an installation's safety zone: authority in this instance would normally be from the installation manager.

Pipe-line activities

Oil or gas which is produced offshore has to be transferred from the installation to an onshore base; this activity is clearly part of the total petroleum activity but, by its nature, the task of transportation is not throughout its duration conducted from or on the installation. Moreover, petroleum extracted in the jurisdiction of one coastal state may well need to be transported across the continental shelf of another state, to be landed in a different jurisdiction from the one in which the extraction was carried out.

Petroleum is in fact often conveyed by pipe-line from the installation to shore, and certainly in the North Sea – because of the deep sea in the Norwegian sector, known as the Norwegian Trough – it is by no means uncommon for petroleum extracted at an installation which is in the Norwegian sector of the North Sea continental shelf to be landed in Great Britain.

It has been necessary in both jurisdictions to make legislative provision to control both the construction and operation of pipe-lines.

British legislation

In providing the framework for the British regulatory system, the Continental Shelf Act made, by S1, installations the focal point of the regulatory system. However it did, in S6, by referring back to S3, make provision for the extension of British law to pipe-lines and cables.

Nevertheless, it was not until the Petroleum and Submarine Pipe-lines Act 1975 that there was any British system for the regulatory control of the construction and operation of pipe-lines. This Act is concerned with the control of construction and use of pipe-lines in British territorial waters and designated areas of continental shelf waters (referred to in the Act as 'controlled waters').

In fact the Act is more concerned with authorisation of the construction and operation of pipe-lines than with the implementation of safe systems of work, although Ss26 and 27 are concerned with these matters.

Section 26 authorises the Secretary of State to make such Regulations as he sees fit to provide for the proper construction and safe operation of pipe-lines, preventing damage to pipe-lines and securing the safety, health and welfare of persons engaged on pipe-line works. Moreover, such Regulations may include provisions with respect to the use of any aircraft, vessel, vehicle, structure, plant, equipment or other thing for the purposes of any pipe-line works.

It is further provided that pipe-line works means works of the following kinds:

(a) assembling or placing a pipe-line or length of pipe-line;
(b) inspecting, testing, maintaining, adjusting, repairing, altering or removing a pipe-line or length of pipe-line;
(c) changing the position of or dismantling or removing a pipe-line or length of pipe-line;
(d) opening the bed of the sea for the purposes of works mentioned in the preceding paragraphs, tunnelling or boring for those purposes and other works needed for or incidental to those purposes;
(e) works for the purpose of determining whether a place is suitable as part of the site of a proposed pipe-line and the carrying out of surveying operations for the purpose of settling the route of a proposed pipe-line.

Section 27 provides for the appointment of inspectors and sets out the duties of such inspectors. The powers of such inspectors include not only the inspection of the pipe-lines themselves, but also the power to enter upon premises, vessels and installations, and to test equipment, if necessary to destruction.

The interpretation of this legislation and the inspection powers which it creates once again involves a careful analysis of the interface between various offshore regimes.

In practice, pipe-laying is likely to be undertaken from pipe-laying barges, which are ships which are subject to flag law – not necessarily the flag law of the country upon whose continental shelf they are operating. However, under S.27 above inspectors are empowered to enter on vessels used in pipe-laying, regardless of their nationality.

In addition, the pipe-line itself must be distinguished from the installation: a pipe-line can extend up to and on to an installation via the riser to any cooling equipment,[9] but S21(10) of the Oil and Gas (Enterprise) Act stipulates that 'installation' does not include any part of a pipe-line. However, a structure on a pipe-line, such as a pumping station, for use in connection with the conveyance of things by pipe-line, will be an installation if it is or will be when established, capable of being manned. This is confirmed by S24 of the Oil and Gas (Enterprise) Act which in sub-section (1)(c) includes places where there is the activity of, 'conveyance of things by means of a pipe or system of pipes . . .' within the definition of installation. Similarly, a gas storage container is, by virtue of S24(1)(b), deemed to be an installation rather than a part of the system of conveyance which is the subject matter of the pipe-lines legislation.

Section 31 of the Petroleum and Submarine Pipe-lines Act excludes from the ambit of the legislation pipe-lines which cross the British sector but have no terminal point in the United Kingdom or controlled waters, except that Regulations may be made concerning them to the limited extent which is permitted by international law. The United Kingdom might, for example, acquire jurisdiction over a particular pipe-line as a result of an agreement with the state(s) where the line terminated. This might possibly occur if there was an international treaty for the joint exploration of a cross-boundary field.

The Submarine Pipe-lines (Inspectors, etc) Regulations 1977 grant power to inspectors appointed under the Act and impose duties on the owners and proposed owners of pipe-lines and others. In particular, Reg 5 requires accidents and other dangerous occurrences to be reported to the Secretary of State for Energy. Regulation 8 provides that none of the inspectors' powers may be exercised so as to impose any obligation on any person on a vessel registered outside the United Kingdom as a ship, aircraft or hovercraft when it is not engaged in operations for the purpose of laying or maintaining a pipe-line.

The Submarine Pipe-lines Safety Regulations 1982 provide for the safe construction, operation (including emergency procedures), periodical inspection and maintenance of pipe-lines. They also require the owner of the pipe-line to give written instructions concerning the movement and anchoring of vessels used in pipe-line work.

The Health and Safety at Work, etc Act 1974 (Application outside Great Britain) Order 1977 in Article 5 extends the Act to pipe-lines within territorial waters or a designated area. The Act's provisions are thus extended to:

(a) any pipe-line works;
(b) the following activities in connection with pipe-line works –
   (i) the loading, unloading, fuelling or provisioning of a vessel,
   (ii) the loading, unloading, fuelling, repair and maintenance of an aircraft on a vessel;
   being in either case a vessel which is engaged in pipe-line works.

This article is especially interesting because it extends onshore legislation offshore, to control the interface between petroleum and shipping activities.

Norwegian legislation

The philosophy of the Norwegian regulatory system takes a broader approach to the extension of domestic Norwegian law across the Norwegian continental shelf than is the case in the British system, with its emphasis on designated areas. It also makes the focus of regulatory control the petroleum activity rather than the installation where the activity is conducted.

These differences of emphasis have enabled Norway to avoid making the detailed provision for construction and maintenance of pipe-lines which has been deemed necessary in the British system. The Introductory Provisions of the principal Act of 22 March 1985 state in S1, that:

This Act applies to petroleum activities ... and pipe-line transportation. The Act also applies to installations for such activities and to shipment facilities for petroleum.

The same section continues to allow (seemingly going further than the British legislation) the King to issue Regulations as to which provisions of the Act shall apply to pipe-lines with associated equipment on the Norwegian continental shelf.

Section 24 spells out that a licence is required to place and operate installations for transportation and utilisation of petroleum:

> Shipment facilities, pipe-lines, liquefaction facilities, facilities for generation and transmission of electricity and other installations for transportation or utilisation of petroleum may not be installed or operated ... without a licence from the Ministry. The licence may be granted for a specific period.

Beyond this, pipe-line systems are dealt with only in broad and general terms in the Regulations concerning safety made under the Act of 22 March by Royal Decree dated 28 June 1985, Chapter VII. Section 49 provides:

> The design, protection, laying, installation, testing and operation of subsea pipe-line systems shall take place in an appropriate manner. Subsea pipe-line systems shall to a reasonable extent, be protected to prevent mechanical damage to the pipe-line due to other activities along the route, including fishing and shipping activities in connection with surveys, or exploration for and recovery of subsea natural resources. Subsea pipe-lines shall be designed, equipped and installed in such a way that they will not cause damage to fishing equipment and that they do not provide an unreasonable obstacle to fishing activities.

The system of internal control is broad enough to enable the licensee to be rendered accountable for pipe-line systems from installations he is operating. The Regulations made by Royal Decree on 28 June 1985 on internal control make no specific reference to pipe-line systems but, on the other hand, Section 6 requires application for approval and envisages that the licensee will submit a general description of the internal control for the total project even before obtaining approval for exploration.

Wide as the Norwegian system of control may be – wide enough to avoid the problems of demarcation between installations and pipe-lines which are encountered in the British system – it cannot avoid the overlaps between shipping and petroleum laws in respect of vessels engaged in pipe-laying activities. However, the facts first, that the Petroleum Act brings vessels within their scope to the extent necessary for control of their petroleum activities and second, that the licensee is required to obtain consent for vessels he employs, enable a wide systems approach to the problem. On the other hand, the emphasis which S54 of the principal Act lays upon the use by the licensee of Norwegian services could be invoked to restrict the use of foreign flag ships in the pipe-laying activity. It is understood that it is not Norwegian policy to impose such restrictions, although economic factors are undoubtedly taken into account in relation to the granting of licences and consents. As a last resort, the coercive powers granted under S65 might be used to remove a foreign pipe-laying vessel from the Norwegian sector of the shelf.

While S1 of the Act of 22 March 1985 describes the scope of the Act as pertaining only to activities (including pipe-line activities) in Norwegian internal waters, in Norwegian territorial seas and on the Norwegian continental shelf, provision is also made for extension of Norwegian law to pipe-lines beyond Norwegian waters and for the extension of Norwegian law to foreign pipe-lines in Norwegian waters:

> The Act also applies to activities and installations ... in areas outside the continental shelf to the extent such application follows from international law or from specific agreements with a foreign state.
>
> The King may issue regulations as to which provisions of this Act shall apply to pipe-lines with associated equipment in areas mentioned in first paragraph, when they are not owned by Norwegian licensees and neither start nor end within the said area.

Diving activities

In both British and Norwegian shelf legislation it was found necessary in the 1970s to make detailed provisions for the control of diving activities on their sectors of the continental shelf. The two countries introduced Regulations which were very similar in style but different from the codes which have been considered so far. It has been seen that the provisions which have been made for most of the situations where there is co-operation between organisations directly engaged in petroleum activities and organisations servicing that activity, have dealt only with the interface between the petroleum and the service activity. The provisions have not attempted to regulate the conditions of service of those in the employment of the sub-contractor responsible for the performance of the service contract: indeed, in many instances the interface has been between shipping and petroleum laws.

The diving Regulations differ in that their main thrust is the safety of the divers themselves: these Regulations are primarily concerned with the working conditions of the divers; the emphasis of the provisions concerning the interface between diving and petroleum activities is the safety of the divers. It is true that for the most part diving will be carried out from a ship alongside the installation but the regulation of the interface between merchant shipping and petroleum laws is not the main feature of diving Regulations.

British legislation

Diving operations in connection with petroleum activities in the British sector were governed by the Offshore Installations (Diving Operations) Regulations 1974, made under the Mineral Workings (Offshore Installations) Act. These Regulations were therefore a part of the regulatory regime applying exclusively to petroleum activities in British territorial waters and on the continental shelf.

The somewhat anomalous situation described above was created by there being a set of Regulations whose focus was on the control of the servicing activity, but whose source was a principal Act whose focus was the petroleum activity. This anomaly has been removed: the 1974 Regulations have been superseded, and diving operations in relation to diving in the context of petroleum activities is now governed by Diving Operations at Work Regulations 1981, made under the Health and Safety at Work, etc Act.

The new Regulations are of wide application: while they apply to, they are not confined to, diving operations on the continental shelf, nor are they exclusively directed towards the petroleum industry. Indeed, it is arguable that the new regulatory system is not part of shelf law in the sense that it is only extended to, rather than made expressly for petroleum activities. It nevertheless provides a good example of legislative provision for safe systems of work. However, it does in fact apply to one of the most dangerous offshore activities.

The importance of the interface between diving and petroleum activities, in those instances where diving is conducted in the context of petroleum activities, is emphasised by Reg 4. This imposes an overall duty on the installation owner, the concession owner (in the case of a proposed installation) and the pipe-line owner to ensure, so far as is reasonably practicable, not only that the Regulations are complied with but that they are complied with in such a way that persons involved are not exposed to risks to their health or safety.

It is also worthy of note that while S5 imposes overall responsibility for the organisation of the system of diving upon the 'diving contractor', who may be the employer of the divers, it also provides that, in respect of diving in relation to offshore installations, in cases where there is no employer the diving contractor will be the installation manager, or the concession owner if it is a proposed installation, or the owner in the case of a pipe-line or proposed line.

The Regulations apply whether the diver is a self-employed or employed person. They impose duties on the diving contractor, the diving supervisor and the divers themselves. They require a system of work to be recorded on paper and its operation to be monitored. They require divers to be qualified, medically fit and to work only a limited number of hours underwater. They also lay down a rigorous system for maintenance, examination and testing of plant and equipment.

The Regulations require that a diving supervisor be appointed in writing: he must be a competent person, either a qualified diver or a person with appropriate experience as a diving supervisor, with adequate knowledge and experience of the diving techniques to be used in the diving operations for which he is appointed. It is his responsibility to ensure, so far as is reasonably practicable, that each diving operation for which he is appointed is carried out in accordance with the diving rules and under his immediate control. Divers are required to have valid certificates of training,[10] a certificate of medical fitness and be competent to carry out safely the work which they are called upon to perform.

At all times when any diving operation is to be carried out there must be present a sufficient number of divers and other competent persons (the 'diving team') to ensure, so far as is reasonably practicable, the operation can be undertaken safely. The Regulations also require that there be one of the specified system of 'stand-by' divers ready to assist the diver(s) engaged in the operation.

The system stresses the importance of documentation. It has been noted that diving supervisors must be appointed in writing, that each diver must have a certificate of training and a certificate of fitness. There must additionally be diving rules, making such provision for the health and safety of persons engaged in the operation as are necessary and specifying the plant and equipment which must be used. The Regulations specify equipment which must be available, including breathing equipment and life-lines, and where the depth of the diving makes this necessary, a diving bell. Plant and equipment must be maintained, examined and tested in accordance with the Regulations. Relevant information concerning the testing and maintenance of the plant and equipment must be recorded in a register kept by the diving contractor.

It is also the diving contractor's responsibility to ensure that every diving operation is carried out from a suitable and safe place with the consent of any person having control of that place. This provision clearly covers the interface between shipping and installation law. If – as is likely to be the case – the operation is to take place from a ship alongside an installation, then the provisions of S5 of the Mineral Workings (Offshore Installations) Act will take effect, to ensure that the final arbiter of whether a diving operation will take place will be the installation manager, if the operation might endanger the safety of the installation or persons on the installation. If the installation manager is also the diving contractor he may have to weigh the safety of the divers against that of the installation. It is possible that he might encounter a conflict of interest, where seeking to ensure the safety of the installation might entail putting the divers at some risk.

Thus, the system under the Diving Operations at Work Regulations ensures that both the diving contractor and the offshore installation manager have responsibilities for ensuring compliance with the Regulations in such a way that persons involved are not exposed to risks to their health and safety. In respect of the installation there is provision that the manager can be appointed as the diving contractor if there is no other person so appointed. In this way there are interrelated duties placed on the manager and the diving contractor to ensure the safety of operations. It is interesting, however, that no mention is made as to the responsibilities of the diving support vessel captain in such matters.

Operations which are conducted from, on, in or near any submersible or supporting apparatus in conjunction with the use of a British-registered vessel are governed by the Merchant Shipping (Diving Operations) Regulations 1975.

Norwegian legislation

Diving connected with petroleum activities in the Norwegian sector of the North Sea continental shelf was governed by the Royal Decree of 9 July 1976 and by Provisional Regulations for diving which were issued by the Petroleum Directorate on 1 July 1978. That these Regulations remain provisional so long after they were issued is apparently the result of the review of the regulatory system which has been undertaken since the new Act of 22 March 1985.[11]

The system established under these Regulations is, however, not dissimilar to that set out in the British Regulations, even though they, unlike the current British Regulations, are solely relevant to diving in the context of petroleum activities. It is not surprising to find that, in the philosophy of Norwegian offshore petroleum laws, responsibility for ensuring that the Regulations are observed rests with the licensee. It is his duty to ensure that a diving contract is only given to an enterprise which has competence in diving work. As is frequently the case with the older Norwegian Regulations, these Regulations contain more detailed standards than do the comparable British ones, which place more emphasis on the duty of those engaged in the operation to exercise their own professional judgement in arriving at safe systems of work.

Interestingly, in the Norwegian system the installation manager has to ensure that communications are effective, which may mean making judgements on the level of human performance rather than on mere technical efficiency of equipment.[12]

It is also interesting that the Maritime Directorate has now issued its own Regulations dated 10 April 1984, concerning the control of diving on Norwegian mobile installations. This new set of regulations would appear to be comparable with the British Merchant Shipping (Diving Operations) Regulations 1975 as, although not entirely comparable in application, both are made under Merchant Shipping legislation rather than under petroleum laws.

The Maritime Directorate Regulations have a requirement that 'diving systems' must be designed and constructed to the IMO Diving Safety Code Standard – a regulatory technique which has not, it is believed, been directly adopted in the comparable British legislation.

Shore base

A good system of communication between the installation and a shore base is a fundamental requirement of safe operation offshore. This must be so for day-to-day operation of the installation: it is, however, of even greater importance in maintaining systems which can be called into operation in the event of an emergency offshore. The consideration of the service activities above, has already identified situations where the regulatory systems require the maintenance of communications between the installation and a shore base. For example, it is of the essence of the telecommunications system and it is also important in the use of helicopters. The identification by the operator of a shore base within the jurisdiction of the country on whose sector of the continental shelf he is operating is also a necessary aspect of the control system of his activities by the coastal state.

The importance of the maintenance of good communications between the installation and a shore base in the context of contingency provisions will be considered in more detail in Chapter 13, which concerns systems for dealing with emergencies.

References

1   For provision for emergencies see Chapter 13; for other communications see Chapter 3.
2   See Chapter 3.
3   See Chapter 13.
4   See Chapter 31 of the *Code of Safe Working Practices for Merchant Seamen*.
5   The failure to ensure that this was so was a cause of loss of life in the Sea Gem catastrophe and it was a cause of concern that many workers were unnecessarily put at risk (all were rescued) in the Key Biscayne incident.
6   See also Chapter 9.
7   Reg 4 of Manning Regulations of 23 March 1982.
8   At the time of writing only an interim report has been published.
9   S33(1) of the Petroleum Submarine Pipe-lines Act 1975.

10 Revised diving training standards have recently been issued. See Health and Safety Information Bulletin No 132 December 1986. See also official Guidance Note *GS 41 Radiation safety in underwater radiography.*

11 See Norwegian Petroleum Directorate *Annual Report* for 1985, p79.

12 Provisional Regulations for diving of 1 July 1978 Reg 2.11.

9

Safe Systems for Plant and Equipment

Principal legislative provisions

British legislation

Factories Act 1961.
Mineral Workings (Offshore Installations) Act 1971.
Health and Safety at Work, etc Act 1974.
The Offshore Installations (Construction and Survey) Regulations 1974.
The Offshore Installations (Operational Safety, Health and Welfare) Regulations 1976.
Diving Operations at Work Regulations 1981.
The Ionising Radiation Regulations 1985.

Norwegian legislation

Act of 22 March 1985 pertaining to petroleum activities.
Act of 4 February 1977 relating to worker protection and working environment.
Regulations concerning safety made by Royal Decree of 28 June 1985.
Royal Decree of 28 June 1985 regulating supervisory activities.
Regulations relating to worker protection and working environment made by Royal Decree of 1 June 1979 as amended 13 September 1985.
Regulations for cranes on production installations of 25 May 1977.
Regulations for production and auxiliary systems on production installations stipulated 3 April 1978.
Regulations for mobile installations with equipment of 10 September 1973 and 13 January 1986.
Regulations on construction, etc of lifts on Norwegian mobile installations of 22 March 1976.
Regulations on deck cranes on mobile installations made 31 January 1978 and 13 January 1986.
Regulations on arrangements above and below deck made on 31 January 1978 and 13 January 1986.
Regulations for drilling of 23 September 1981.
Provisional Regulations for diving stipulated 1 July 1978.
Regulations of 10 April 1984 concerning the control of diving systems.
Regulations for electrical installations on drilling platforms made 1 December 1974.
Temporary Regulations for electrical equipment on mobile installations of 26 July 1985.

In a previous chapter emphasis was placed on the use within regulatory systems of the process of independent scrutiny for evaluation of the design and construction of primary structures used in the offshore oil industry, at or before the commencement of petroleum activities. An obvious limitation to this technique as a mechanism for monitoring the installation and maintenance of safe plant and equipment is that, after the initial survey, the structure or its equipment may be modified or submitted to stresses which ought to invalidate the original assessment. Additionally, even if there are no overt changes made which might invalidate the assessment, constant inspection and maintenance by the operator will be required to keep the installation in an acceptable condition. The certifying authority must be notified in accordance with Reg 7(3)(c) of the Offshore Installations (Construction and Survey) Regulations if plant and equipment is changed after certification.

These examples serve to illustrate the limitations of the certification and approval processes. It is clear that areas will remain where safety offshore will require the imposition upon the operators of some type of direct legal responsibility for the inbuilt safety of the mass of equipment used offshore and, equally importantly, for its regular maintenance.

Onshore safety provisions

The types of responsibilities mentioned above are analogous to those imposed upon the management of onshore industrial enterprises, found in legislation like the British Factories Act 1961. In this Act, the statutory formula for regulating plant equipment and machinery which is not dangerous in the course of normal working has been refined over the years. It now consists of the classic language, 'good mechanical construction, sound material and adequate strength and be properly maintained'.[1]

This formula (or its derivatives) is nowadays applied to, *inter alia*, hoists and lifts, cranes and lifting machines, steam boilers and pressure systems. It has been interpreted by the courts as imposing an absolute duty: not only an initial duty to ensure that the plant is of good construction, but also a continuing duty to consider whether it has reached or passed the limits of its safe working life. Arguably, a misuse of otherwise adequate equipment – for example, the overloading of a crane which leads to failure – is not a breach of this requirement, but would probably amount to a breach of some other substantive requirement of the safety code.

'Maintenance' is defined in the same legislation as: 'maintained in an efficient state, in efficient working order and in good repair'.[2] Bearing in mind the strict nature of the basic plant and equipment obligation, it is perhaps not surprising that when the maintenance obligation is read into this formula, the courts have taken the view that where a machine fails in an abnormal situation an offence is committed, despite evidence of an efficient system of routine inspection and maintenance. In other words, the law describes by the phrase 'properly maintained' the result to be achieved rather than the routine maintenance to be adopted.

Where equipment is normally dangerous, in the sense that it is dangerous to an operator even when it is of optimum construction and maintenance (for example, power presses), British onshore safety legislation has for many years made special provision for protecting workers from the normal hazards of the machine. Thus, every dangerous part must be securely guarded. In this instance the emphasis is on normality; the legal formula, which gives rise to a strict duty, is nevertheless satisfied by a device adequate to protect against the hazards encountered in normal working and need not safeguard against abnormal failures such as a machine which disintegrates while in rotation.[3] Hazards of this latter type are identified in the construction and maintenance formulae already discussed.

Offshore legislation

Although these onshore safety provisions have no immediate application offshore, the long experience of enforcement which they encapsulate has been drawn upon in formulating offshore legislation: indeed, much of this statutory language can be identified in the British offshore legislation.

The same problems obviously faced other countries while they were drafting their offshore legislation and similar thinking can be identified in the Norwegian legislation.[4] It is important to recall, however, that the Norwegian system relies upon prior approvals and consents of all descriptions to a far greater degree than does the British system, while both systems have relied upon the certifying authorities. In Britain the only prior approvals required are for those structural matters discussed in Chapter 6. In Norway, on the other hand, as a matter of normal practice until recently little could be initiated by management without the prior approval of the inspectorates onshore and offshore.[5]

However, where maintenance as opposed to provision of equipment is concerned, systems of prior approval can only operate upon a paper system of maintenance and can be no guarantor for the subsequent performance of that system. Where maintenance and allied matters are concerned, it appears that in all jurisdictions there can be no substitute for performance obligations imposed upon management, and enforced by some system of inspection.

The problems which have to be considered are therefore: 1) ensuring that plant and equipment is initially of an appropriate standard; and 2) ensuring that both plant and equipment are adequately maintained.

Plant and equipment standards

British legislation

Construction and survey requirements

The standards for equipment forming part of an installation at the time of 'establishment' or 'bringing into' the British sector of the continental shelf of the

North Sea are to some extent included in the subject matter for certification and approval under the Construction and Survey Regulations. Schedule 2, Part VIII sets out the 'General Requirements' which must be complied with as a condition precedent to certification and approval, including fairly detailed provisions as to equipment generally:

1. Every item of equipment –
   (a) shall comply with a recognised standard or specification;
   (b) shall be suitable for its intended purpose and incorporate efficient control apparatus, guards, shields and other means of protecting personnel;
   (c) shall be located to ensure safe operation and, if located in an area within which danger of fire or explosion from ignition of gas, vapour or volatile liquid exists or is likely to exist, shall be suitable for use in that area;
   (d) shall be provided with a safe means of access;
   (e) if capable of causing noise or vibration which is, or is likely to be, injurious to health, shall be suitably insulated; and
   (f) shall be so installed and disposed, both individually and in relation to other items of equipment on the installation, as to reduce to a minimum any potential danger to the installation and its personnel.

However, experience of onshore safety legislation and its enforcement leads to the conclusion that these safety standards for equipment, being essentially non-specific and not intended to be the subject matter of a criminal contravention, might prove difficult to enforce in the case of, for example, a dispute concerning the integrity of a machine or device intended to be installed offshore. This is not only on the grounds that the Regulations are intended to deal only with equipment initially provided (and its replacement), but possibly not other equipment acquired later: even in the case of initially provided equipment, the only sanction laid down by the Regulations in case of dispute, is refusal to grant the necessary approval – a decision from which no appeal is provided. It is most unlikely that the approval of the whole installation would be held up over, say, a disputed machine: it is therefore possible that a certificate might be issued with conditions relating to the use or non-use of the plant or equipment in question.

Operational requirements

The Mineral Workings Act 1971 contains in S6 a wide power to make Regulations, although on first reading the section appears to be placing more emphasis upon safe systems of working than upon the provision of equipment. However, a perusal of paragraph 4(1) of the Schedule to the Regulations, which serves to amplify S6, makes it clear that Regulations covering equipment can indeed be made:

… as to the equipment, facilities or materials which are to be, or may be supplied or used, whether the provision has reference to sufficiency, to suitability, to safety during use or while not in use, or to any other matter …

Thus, this power to regulate any equipment is in reality quite comprehensive. In the implementation of this power, the Operational Safety, Health and Welfare Regulations 1976 include, in Reg 10, requirements that all equipment of an offshore installation shall be of good construction, sound material, adequate strength, and free from patent defect, and suitable for any purpose for which it is used. It should be recalled that each of these provisions is capable of constituting the basis of a substantive offence.

It is clear that the inspectorate now possesses a power to enforce a requirement for alteration or improvement of any item of equipment used, or proposed to be used, offshore. The certification procedure would, almost by definition, fail to be as effective in this respect since it depends on the initiative of the operator and acceptance of the certifying authority.

In marked contrast to certification procedures, a prosecution based upon Reg 10 of the Operational Safety, Health and Welfare Regulations would be conducted in an onshore court, which would rely upon the great deal of learning accumulated around the similar formulae under the Factories Act and other safety legislation. It is interesting to speculate as to the scope of the phrase: '... suitable for any purpose for which it is used ...'

It has already been pointed out that wrongful use leading to the failure of a machine does not of necessity constitute a breach of the equivalent onshore provisions. However, the incorporation of this phrase into Reg 10 raises the possibility that, in the instance of the failure of a crane offshore through gross overloading, the owner or the offshore installations manager might be vicariously liable for a breach of this Regulation.

It is significant that when the same Regulations deal with the safeguarding of equipment used offshore (ie equipment that is for some reason dangerous in the course of normal working), current statutory language in use onshore has again been borrowed, but with some significant departures. Thus, the duty in Reg 12 to 'efficiently guard' any dangerous part of any machinery or apparatus is qualified by 'practicability' – no doubt in deference to the need to ensure adequate operational machinery offshore – while other qualifications upon the duty to guard incorporate a great deal of onshore experience. Thus, the Regulation provides for both static and movable (ie interlock) guards, and makes a concession for the removal of guards for 'examination, adjustment, or lubrication', provided that certain specified steps and practical arrangements are made to prevent injury while these operations are being carried out.

Electrical equipment is dealt with separately under Reg 11. This specifies that all electrical equipment shall be sufficient for the work for which it is to be used, and so constructed, installed, protected, worked and maintained as to prevent danger '... so far as is reasonably practicable'. It appears that Reg 11 leans heavily on the Electricity Regulations applicable to onshore industry, Reg 1 of which states:

> ... All apparatus and conductors shall be sufficient in size and power for the work they are called upon to do, and so constructed, installed, protected, worked and maintained as to prevent danger, so far as is reasonably practicable.

The Operational Safety, Health and Welfare Regulations do acknowledge the the problems of multi-employer situations on installations, at least as far as electrical apparatus is concerned. Regulation 17(2) is as follows:

> No person shall take on to, or use on, an offshore installation any apparatus designed

for the generation, conversion, storage, transmission, transforming or ultilisation of electricity, which is not to be used as, or form part of, the equipment of the installation, except with the written permission of the offshore installation manager.

It is useful to compare this provision with the language of Reg 10 which relates to the integrity of the 'equipment of an installation', and Reg. 12 relating to the safeguarding of 'any machinery or apparatus'. Although a court might well rule that Reg 12 covered equipment belonging to whomsoever, Reg 10 might well be restricted to equipment forming part of the installation. Regulation 17(2) on the other hand, clearly contemplates contractors' equipment that is not intended to form part of an installation; and for that reason introduces the principle, novel to the Mineral Workings Regulations, of prior approval by the installation manager before the introduction or use of such equipment.

Detailed requirements for the provision of special equipment will be found in other Regulations made under the Mineral Workings (Offshore Installations) Act, and important in this context are the special provisions for cranes and lifting tackle which are contained in Reg 6. This Regulation and Part III of Schedule I of the Regulations provide for a tight system which applies to 'every lifting appliance, and every piece of lifting gear', requiring it to be examined and, where necessary, tested by a competent person (who is neither the owner of the installation nor his employee). These procedures are to be followed:

(a) before it is used for the first time; or
(b) having already been used, if and whenever subsequently substantially altered or repaired, before it is again used; and
(c) at the times and intervals set out in Part III of the Schedule ...

which is a thorough examination every six months.

Few additional Regulations have been made as to the construction or safeguarding of equipment, other than the approvals system described in Chapter 6, the Offshore Installations (Life-saving Appliances) Regulations 1977, and the Guidance for Life-saving and Fire-fighting Equipment. Otherwise, such matters must generally fall under the Operational Safety, Health and Welfare Regulations discussed above.[7]

The Health and Safety at Work, etc Act 1974

The application offshore in 1977 of the British Health and Safety at Work, etc Act 1974 was of considerable significance in the context of ensuring safe plant and equipment; for both the general duties and the regulation-making powers are most relevant to both the control and maintenance of plant and machinery.

Under S2(2) of this Act an employer is under a duty to ensure:

... the provision and maintenance of plant and systems of work that are, so far as is reasonably practicable, safe and without risks to health.

Plant is defined as including 'any machinery, equipment, or appliance'.

Despite the sweeping nature of this provision, it has been specifically enforced by the courts on many occasions. Although at first sight the defence of 'reasonable practicability' appears to impose a standard lower than that imposed by the language of the Operational Safety, Health and Welfare Regulations already discussed, the defence, as elaborated in S40 of the Act, is quite narrow. Any defendant wishing to invoke the defence must himself prove that it was not reasonably practicable to do more than was in fact done to satisfy the duty or requirement. In practice, it appears difficult successfully to invoke the defence and satisfy a court as to the reasonableness of one's own conduct in responding to any of the requirements of the Act or Regulations.[8]

It has already been noted that the Act imposes duties upon employers and managers for the benefit not only of their own employees, but also those of other employers who are at risk from the activity in question.[9] Yet another novel requirement is found in S6, which imposes duties upon designers, manufacturers, importers and suppliers of 'articles for use at work'. Such persons are under a duty to ensure, so far as is reasonably practicable, that the article is so designed and constructed as to be safe and without risks to health when properly used; to test, or arrange for the testing of, such article as may be necessary to comply with the general duty outlined above; and to take steps to secure the availability 'in connection with the use' of such article, of adequate safety information necessary to ensure its safe use at work. 'Proper use' is defined in relation to use in compliance with the safety information supplied with the article in conformity with these provisions. 'Article for use at work' is defined as:

> any plant designed for use or operation (whether exclusively or not) by persons at work, and
> any article designed for use as a component in any such plant.

It would be difficult to formulate a wider definition of article, or to identify a device used at work which would not fall within it, other than a 'substance' such as a raw material. Thus its application offshore is very widespread indeed; the only limitation, probably, being the installation structure, as opposed to the components, machines, and equipment that rest upon or are secured to the structure of the installation.

Although S6 was primarily drafted with onshore conditions in mind, there can be no doubt that its application offshore cannot fail to have a beneficial effect. Much offshore equipment is of great complexity and specialisation and, wherever it may be designed, and wherever it may be manufactured, it must be supplied to meet British safety standards, and not merely those of the country of origin. So far no prosecutions have been brought under this provision offshore, but this can be no measure of its effect upon offshore hazards.

In the Norwegian system, on the other hand, comparable standards are achieved through monitoring of the licensee's audits (although the police may prosecute independently of petroleum laws).

The regulation-making powers of the Health and Safety at Work, etc Act contained in S15 and Schedule 3 make no special provision for a 'prior approvals' system. There is a licensing power[10] - which, however, has been seldom used - but there is a most comprehensive power to regulate plant and equipment already in use.[11] Regulations may be made:

> ... imposing requirements with respect to the design, construction, guarding, siting, installation, commissioning, examination, repair, maintenance, alteration, adjustment, dismantling, testing, or inspection of any plant.

Regulations made under these powers may vary in their degree of strictness. While most codes of Regulations issued since 1974 have been qualified by the 'reasonably practicable defence', there is no reason why, in appropriate cases, plant and equipment Regulations should not be made, adopting the same statutory language and importing the same degree of strictness as those made under the Factories Act 1961; and no reason why such strict Regulations should not be applied offshore.

It is interesting to note that Reg 12(4) of the Diving at Work Regulations does make provision for a safe system for plant and equipment used in diving:

All plant and equipment used in a diving operation shall –
(a) be properly designed, of adequate strength and of good construction from sound and suitable material;
(b) be suitable for the conditions in which it is intended to be used;
(c) where its safe use depends on the depth of the pressure at which it is used, be marked with its safe working pressure or the maximum depth at which it may be used;
(d) at whatever temperature it is to be used, be adequately protected against malfunctioning at that temperature.

In addition to penal sanctions available under the 1974 Act, enforcement notices may be served by inspectors to secure safe systems in plant and machinery cases offshore. On the basis of the experience of the use of notices onshore, one may predict that both improvement and prohibition notices could be used extensively offshore in plant and machinery cases. In appropriate cases, the continued use of the offending equipment may be prohibited.

Any differences of opinion between management and inspectors over the legality of the matter in dispute will have to be resolved at a later date after the issue of the notice, and the dispute cannot be used, as under other enforcement procedures, as a mechanism for delay. Where a real risk is perceived by the inspector from the continued use of the equipment, he will serve an immediate or deferred prohibition notice.

Regulations made under the Health and Safety at Work, etc Act only apply offshore if they have been formulated with express provision for application on the continental shelf. The Diving Operations at Work Regulations referred to above, are an example of Regulations which have been made to apply to petroleum activities. Additionally, the following Regulations have been so applied:

Control of Lead at Work Regulations 1980.
Asbestos (Licensing) Regulations 1983.
Freight Containers (Safety Convention) Regulations 1984.
Asbestos (Prohibition) Regulations 1985.
Ionising Radiations Regulations 1985.

No doubt we shall see in the future other examples of specific construction and maintenance standards incorporated in Regulations that are specifically applied offshore.[12]

Norwegian legislation

For largely historical reasons, Norwegian legislation for fixed installations is still somewhat different from that for mobile installations, although in the future, when the rationalisation intended by the Act of 22 March 1985 is completed and now that shipping and petroleum activities are separated, the historical distinctions are likely to fade away.

Already the principal Norwegian legislation governing plant and equipment standards offshore is the Royal Decree of 28 June 1985, laying down safety Regulations to supplement the principal Act on petroleum activities. This applies to both fixed and mobile installations.

As a framework Decree, its provisions are as general as those in the British Health and Safety at Work, etc Act. Thus, Reg 19 calls for 'testing and control' by the licensee before the use of 'installations'; while Reg 20 requires 'testing and inspections' during operation of installations. By Reg 21, 'installations and equipment' shall at all times be in a 'proper condition'. Electrical installations are required to be 'designed, used, and maintained properly'[13] and 'lifting arrangements' must be 'soundly constructed and equipped'.[14]

What is perhaps more significant under the Norwegian system, is that Regs 8–12 deal with the power of the regulatory authorities to issue consents – without which no petroleum activities can take place. Of most significance for the use of plant and equipment are the powers set out in Reg 11. For example, consent must be obtained before the 'initiation of detailed engineering', prior to 'initiating fabrication'; and before 'an installation or parts of it are put into service'. Applications for approvals and consents as required by Regs 8–12 must, by virtue of the Royal Decree of 28 June 1985 concerning regulatory supervisory activities, be submitted to the Petroleum Directorate.

The other principal legislation dealing with plant and equipment offshore is the Act of 4 February 1977 relating to worker protection and working environment onshore, currently applied to fixed installations offshore, in part, by Royal Decree of 1 June 1979. This Act is more specific than the Royal Decrees discussed above, but relies upon the same system of prior approval of equipment and activity by the inspectorate.

Section 9 of the Act provides for technical apparatus and equipment in the following terms:

1 Technical apparatus and equipment in the enterprise shall be designed and provided with safety devices so as to protect employees from injury and disease.
   When technical apparatus is being installed and used, care shall be taken to ensure that the employees are not exposed to undesirable effects from noise, vibrations, uncomfortable working positions, etc.
   Technical apparatus and equipment should be designed and installed so that it can be operated by, or be adapted for use by, employees of varying physique.

The section then makes provision for Regulations to be made by the King concerning the requirements imposed under the section, including:

(a) design, construction, installation, etc.
(b) approval.

(c) tests of materials, or the examination or inspection of technical apparatus and equipment by experts.

Under S10, steam pressure systems must be properly manufactured, installed, equipped and maintained. They must not be used without the prior permission of an inspector. Regulations may be made relating to inspection and maintenance. The system, which applies initially to steam pressure systems, may be extended to other pressure systems by regulation.

Multi-employer situations are regulated under S15. Each employer in such a situation is under a duty to safeguard the employees of others, and under a duty to co-operate with the other employers involved for this purpose. The principal employer in this situation has the co-ordinating duty. Section 17 imposes a duty upon manufacturers of technical apparatus or equipment 'which will, or may foreseeably, be used by enterprises covered by this Act', to ensure that they are designed and provided with safety apparatus so as to comply with the Act. Such equipment must be provided with safety instructions in Norwegian, and provision is made for Regulations relating to design, construction, installation and approval of all such equipment.

The powers of inspectors under this Act are set out in Section 77. Appeals against refusals and decisions of individual inspectors may be taken to the Directorate of Labour Inspection.

Detailed Regulations
More detailed construction provisions are found in the Regulations made under these framework ordinances. For instance, detailed requirements as to equipment will be found in the Regulations of 25 May 1977 relating to cranes on production installations. Although there is much detail concerning plant and equipment, most of these are what might be termed minimum requisites, rather than construction or performance standards. There are requirements for 'thorough examination' of equipment (which incidentally, is defined as 'an examination that is so thorough as to permit a reliable judgement regarding the safety of the parts examined and tested'). Testing, examination, and certification is dealt with in Reg 2.8: this requires that each crane shall have a certificate relating to the characteristics and history of the crane; this, together with details of routine tests, must be entered in a crane control book.

Similarly, production and auxiliary systems Regulations of 3 April 1978 contain much detail as to equipment requisites and systems of operation, but also a general requirement that pressure systems and other equipment which 'otherwise has an influence on safety' shall be 'designed, fabricated, installed, and tested' in accordance with the Regulations.[15] The materials of which such systems and equipment are constructed must be of 'adequate strength, toughness, ductability, homogeneity, and durability'.

There is a lack of provision to ensure the adequacy of equipment, but it must be borne in mind that all equipment must meet with the approval of the agency prior to its installation or use. In these circumstances the lack of express regulatory standards may be less significant.

There appears to be somewhat more detailed regulation of navigational plant and equipment applicable to Norwegian mobile drilling units. However, it must be recalled that the legislation set out in the 1986 edition of the *Red Book* applies only to Norwegian-registered mobile drilling units and thus forms part of what might be properly regarded as Norwegian merchant shipping law.[16] Some Regulations – as listed in the Description of Supervisory Activities of 28 June 1985 and set out in Volume I of the Petroleum Directorate's publication – apply to all mobile installations operating on the Norwegian continental shelf. They may therefore properly be regarded as forming part of the Norwegian petroleum legislation, enforceable against whomsoever is engaged in petroleum activities in the Norwegian sector of the continental shelf. Again, most relevant Regulations concern requisites, rather than construction and maintenance standards; and again, much emphasis is placed upon the consent system.

Regulations for mobile drilling platforms with installations and equipment dated 10 September 1973 (shelf) – Regs 3-6 of which deal with applications for consent to use a mobile drilling unit – contain detailed provisions for mobile installations. Regulation 9 provides for inspection and maintenance programmes for anchoring systems. Regulation 20 requires all life-saving and emergency equipment to be maintained in good order; it also includes a list of emergency equipment that must be examined and tested by a competent person. Regulation 21 requires that all mobile drilling units must keep an equipment maintenance record containing details of all tests prescribed by Reg 20, including 'details of periodic maintenance', in the form specified in the Regulation. The Inspectorate may withdraw consent for use of the mobile installation for 'serious or repeated' violations of these provisions. Comparable flag Regulations were issued on 13 January 1986.

Regulations for lifts on Norwegian mobile drilling units and ships of 22 March 1976 lay down in Reg 6 that lifts shall be constructed of solid materials and have adequate strength under all conditions of operation. Regulations on deck cranes on mobile drilling units dated 31 January 1978 (shelf) and 13 January 1986 (flag) require, in Reg 8, that a crane manual be kept on board, with information on design and construction. They also provide rules for maintenance and inspection, while Reg 11 provides for testing of cranes.

Regulations on arrangements on and below deck dated 31 January 1978 (shelf) and 13 January 1986 (flag), lay down construction standards for stairs and fixed ladders, and in Reg 8 there is a requirement that 'projecting moveable parts' which are not shielded by position or construction, must be provided with 'adequate protective arrangements'.

Under the provisional diving Regulations of 1 July 1978, applicable to all diving in connection with petroleum activities on the Norwegian continental shelf, the diving contractor (who under Reg 13(2) must provide all necessary plant and equipment) must secure that that plant and equipment is of sound construction and suitable material, in good working order at all times, and adequate for the purpose for which it is used. Under the same Regulation, other equipment, for example compression chambers and diving bells, are to have a 'sound construction standard', while other parts of diving equipment shall 'fulfil accepted standards for material strength'. Regulations 14 and 15 provide for testing of equipment at specified intervals, together with 'daily examinations'.

Electrical equipment on mobile drilling units is regulated by temporary

Regulations dated 26 July 1985 (shelf), which provide for survey, approvals and withdrawals of permission to use electrical equipment where there have been 'serious or repeated' violations.

Drilling is covered by Regulations of 23 September 1981 which, again, apply to drilling from any type of installation. Under Reg 2.10, all equipment used during drilling operations must be designed, manufactured, installed and tested in accordance with 'recognised or accepted standards'. This phrase can be seen as another manifestation of 'good oil field practice', which have been translated, through the voluntary codes of the American Petroleum Institute and other bodies.[17]

Maintenance standards

British legislation

A careful reading of the Construction and Survey Regulations does not support an assertion that standards relating to the maintenance of plant and equipment offshore could be enforced through the mechanism of these Regulations. While the equipment that might form part of the subject matter of a certificate of fitness means:

> ... any plant, machinery, apparatus, or system attached to or forming part of an offshore installation,

it is not clear how maintenance conditions might be attached to its approval, except possibly through the mechanism of the 'operations manual'. An application for a certificate of fitness for a fixed or mobile installation must be 'accompanied' by the operations manual relating to the installation. This manual is defined as: written particulars provided by the owner of an offshore installation for the 'information, guidance and instruction' of the manager, relating to the 'safety' of the fixed installation when on station, and the 'seaworthiness and stability' of a mobile installation. There is no provision within the Regulations for the specific approval of the manual, although this is implied by Reg 5(2) which provides for an 'independent assessment' of the manual, to ascertain whether the information, guidance, and instructions therein are 'adequate and appropriate' in relation to the installation. The certifying authority has to check the receipt of the manual, whether there has already been an independent assessment commissioned by the owner or not.

While the processes of survey and assessment of the application for a certificate of fitness are being carried out, no alteration may be made in any of the provisions of the operations manual. No other reference is made to the manual, leaving unresolved the questions: 1) are the contents mandatory?; and 2) if so, could a certificate of fitness be terminated under Reg 11 for failure to follow the precepts within the manual? These issues are only germane to the question of maintenance of equipment if systematic maintenance were a requisite of the manual.

It is difficult to read such requirements into the provisions outline; in one sense, proper maintenance clearly relates to the safety of an installation, but difficult indeed to read such a requirement into 'seaworthiness and stability'. In fact, an interpretation of these terms that met all these requirements, and for that matter fitted into the concept of 'operations', would be instructions on how to make use of equipment and fittings, rather than how to maintain its efficiency and safety by maintenance.

Even if this argument is incorrect, and maintenance does form part of the mandatory requirements of an operations manual, it still leaves unresolved the second question relating to its enforcement. Assuming that the maintenance of equipment on an installation was not carried out in conformity with the manual, what sanctions under these Regulations could be applied? The only possibility would appear to be Reg 11(1)(b), under which the Secretary of State might terminate a certificate of fitness if: 'the installation to which it relates is not, or is no longer, fit to be maintained in the relevant waters ...'

It might possibly be argued that an installation with, for example, poorly maintained cranes, is 'unfit' to be maintained in relevant waters; but the argument lacks conviction and would, it is submitted, fail to stand up in court if a challenge were brought against the inspectorate by the installation owner. In fact in seems unlikely that the Construction and Survey Regulations would be the appropriate vehicle for the enforcement of maintenance standards offshore.

The powers contained in the Mineral Workings Act seem much more adequate to regulate standards of maintenance. Thus the regulation-making powers under S5 and the Schedule include, 'measures to ensure the safety of the installation', and provisions as to the 'equipment' of any installation. Under these powers the Operational Safety, Health and Welfare Regulations 1976 were made, which contained, in Reg 5, a general maintenance requirement which states that all parts of every offshore installation and its equipment shall be so maintained as to ensure the safety of the installation and the safety and health of the persons thereon.

This obligation is reinforced by Reg 5(2) which requires the maintaining in force at all times of a scheme for systematic examination, maintenance and testing of 'all parts of an installation and its equipment'. The scheme must specify the intervals within which such examination and testing must take place, together with a specification of the requisite test and examination in each case. The minimum intervals at which certain sensitive equipment must be examined and tested are set out also in the Schedule to the Regulations. The scheme of maintenance, together with the records of the examination and testing carried out under it, must be preserved for a minimum period of two years.

It is interesting to compare these requirements with those developed for maintenance in onshore industry, as set out in the current Factories Act. There the requirement for 'proper maintenance' has been interpreted to create a strict duty; so strict that it relates to the results, rather than the process, of maintenance. Offshore, the object of the maintenance exercise has been clearly set out; namely, ensuring the safety of the installation. It will be interesting to observe how the courts will interpret this requirement: either as steps that should, in normal circumstances, secure the result, or, on the onshore analogy, as steps that will by definition be inadequate unless the result is achieved – so that a failure is treated as a contravention.

In any event, Reg 5 has gone on to systematise the maintenance requirement by specifying the system that must be set up for monitoring the maintenance arrangements both generally, and, in specific instances, on the basis of a statutory formula. The record-keeping requirement is also significant as forming part of the classic 'enforcement by inspection' technique; the inspector, on his visits not only monitors the present state of the equipment, but can assess it against the records of previous maintenance procedures and their implementation.

The provisions of the general duties under the Health and Safety at Work, etc Act are relevant to the question of maintenance standards offshore. Section 2(2)(a) specifically refers to the 'maintenance' of plant and systems so far as is reasonably practicable. The definition of plant under this Act, it will be recalled, is wide enough to encompass all the equipment of an installation. There has been little case law on the extent of the duty to maintain equipment under this legislation; but it may be said that the courts are prepared to spell out criminal contraventions from conduct which can be established as in breach of the broad and sweeping obligations imposed by the general duties. But it seems likely that, given the detailed nature of the maintenance obligations imposed by the Operational Safety, Health and Welfare Regulations, the inspectorate will continue to rely on the latter when concerned with lapses of maintenance offshore.

Specific provision for maintenance can also be found in Reg 13 of the Diving Operations at Work Regulations, made under the 1974 Act. The particular items of equipment whose use is mandatory under these Regulations may not be used unless they are maintained in a condition which will ensure, so far as is reasonably practicable, that they are safe while being used.

This general requirement, which on the face of it is not as strict as the general maintenance provisions already discussed, is supported by more detailed obligations for examination of the equipment at specified times by a competent person, who must certify in writing that it is safe to be used. All certificates relating to these tests must be preserved by the diving contractor for a period of two years, except in the case of compression chambers and diving bells, where the records must be preserved for five years.

An interesting maintenance obligation is introduced offshore by the Freight Containers (Safety Convention) Regulations 1984. Containers to which the Regulations apply must be regularly examined in accordance with a procedure that has been approved by the Health and Safety Executive and the Department of Energy.

Norwegian legislation

The Norwegian system of prior consents, which imposes a great deal of documentation as to the state of equipment and machinery at the moment of installation has, almost by definition, little inspectoral involvement in routine operations such as maintenance. Here, only a legal obligation, analogous to British legal procedures, which places ongoing legal responsibility upon management or operators, can be effective, and this is achieved through the concept of internal control. Under the system now in force the licensee has a general duty to conduct internal audits to ensure that all regulatory provisions are observed and standards maintained.

From time to time in the Regulations quoted above there have been references to 'examination, testing, and maintenance'. Where these have to be systematised and recorded in advance, the consent system pronounces upon the adequacy of the operator's proposals for maintenance. Where records have to be kept, this is evidence available for inspectors when they pay routine visits to check the licensee's audits; and, of course, investigation following incidents or accidents can be used to monitor the working of a maintenance system.

It might be observed that the recent increased reliance in Norwegian legislative regulation upon the operators' system of 'internal control'[18] is entirely in sympathy with the need for safe systems, as it is precisely in areas such as maintenance that the move away from approvals to operator responsibility can be seen to follow a logical progression.

References

1 Eg S22: every hoist or lift shall be of good mechanical construction, sound material and adequate strength, and shall be properly maintained.
2 S176.
3 Eg *Close* v *Steel Company of Wales* [1962] AC 367.
4 See, for example, the State of Western Australia's *Direction as to Safe Practice – General* (dated 1 August 1979). While it is not clear that this is more than a code of practice, it makes provision for guards and fencing of plant and machinery used on the relevant continental shelf.
5 The system of 'consents' now operated relaxes the direct control of the Norwegian Petroleum Directorate somewhat.
6 Schedule 2, Part VIII.
7 Apart from approvals for fire-fighting and life saving appliances.
8 The absence from the law reports of prosecutions means guidance on judicial attitudes can only be obtained by reference to reports of appeals against improvement and prohibitions notices (Ss21–25).
9 See Chapter 7.
10 Schedule 3, paragraph 4.
11 Ibid, paragraph 1(2).
12 It is most unlikely that Factories Act provisions can be applied offshore directly: however, they and onshore provisions made under the Health and Safety at Work Act could be invoked to demonstrate what is 'reasonably practicable'.
13 Reg 38.
14 Reg 41.
15 Reg 3.1.
16 Compare first British model clauses. See Chapter 2.
17 Eg Regulations of 31 January 1978 on deck cranes on mobile installations.
18 See Chapters 4 and 5.

10

Safe Systems: Manning, Qualifications and Training

Principal legislative provisions

British legislation

Mineral Workings (Offshore Installations) Act 1971.
Health and Safety at Work, etc Act 1974.
Merchant Shipping Acts 1894–1984.
The Offshore (Operational Safety, Health and Welfare) Regulations 1976.
The Offshore Installations (Emergency Procedures) Regulations 1976.
Diving Operations at Work Regulations 1981.
The (Proposed) Offshore Installations and Pipe-line Works (First Aid) Regulations.

Norwegian legislation

Act pertaining to petroleum activities of 22 March 1985.
Regulations to implement the principal Act made by Royal Decree of 14 June 1985.
Regulations concerning safety by Royal Decree of 28 June 1985.
Worker Protection and Working Environment Act of 4 February 1977.
Licensee's internal control by Royal Decree of 28 June 1985.
Control of foreigners by Royal Decree of 25 August 1967.
Regulations relating to worker protection and working environment made (amending RD of 1 June 1979) by Royal Decree of 13 September 1985.
Provisional guidelines for the training and furnishing of certificates for crane operators on mobile drilling units and other mobile installations of 23 November 1978.
Regulations for mobile drilling platforms of 10 September 1973 and 13th January 1986.
Regulations on certificates of competency for personnel on drilling vessels and other floating installations of 11 December 1981.
Regulations concerning the manning of Norwegian drilling units and other mobile installations of 23 March 1982.
Regulations concerning qualification requirements for personnel on drilling units and other mobile installations of 23 March 1982
Provisional Regulations for diving stipulated 1 July 1978.
Regulations concerning working hours for diving personnel on drilling platforms of 26 June 1981.

Regulations concerning the control of diving systems on mobile installations of 10 April 1984.

Regulations for cranes on production platforms of 25 May 1977.

Regulations for cranes on mobiles of 31 January 1978 and 13 January 1986.

Regulations concerning medical examination of employees of 1 August 1980.

Provisional standard directive for SRN of 11 April 1976.

Standard instruction for physicians in charge of fixed installations May 1982.

Regulations on life-saving appliances of 3 February 1982.

Regulations for drilling of 23 September, 1981.

Regulations concerning qualification requirements of drilling personnel of 22 February 1983.

Regulations on aeromobile equipment on mobile units of 8 September 1980.

Regulations concerning aeromobile equipment on production platforms of 10 September 1980.

Regulations for helicopter decks of 18 April 1983.

Safe systems of work are not only a question of the provision and maintenance of safe plant and equipment. Safety depends on the selection of personnel in sufficient numbers and with sufficient qualifications to undertake tasks safely, including the ability to operate plant and equipment.

Training is an important factor in the maintenance of safe systems. Personnel may be qualified at the outset of their contract of employment, but it is still likely that they will require in-service training in order to retain their original skills, and to keep abreast of changing work systems.

Additionally, it is not sufficient merely to install safe systems: persons, like plant and equipment, must be monitored to ensure that the systems which have been adopted are actually adhered to throughout the performance of the task. Indeed, the systems for maintaining plant and equipment and recording that testing and maintenance have been undertaken, are really systems for monitoring the performance of the personnel entrusted with the responsibility of maintaining the equipment.[1]

Regulatory legislation which seeks to achieve safe systems of work will need to make provision for ensuring that people operate safely, as well as ensuring that they are provided with suitable plant and equipment. Legislation may approach this problem in varying degrees of detail. At one end of the spectrum, legislation may merely impose upon the organisation a very general responsibility to maintain a safe system, and leave to that organisation the task of identifying the way to achieve this objective. At the other extreme, legislation may stipulate in detail the systems to be adopted and the qualifications and training required by the personnel who are to be employed in these tasks. A midway stance is for the legislation to identify the tasks which have to be carried out, and the situations provided for within an organisation, and leave it to the organisation to devise its own means and detailed criteria for achieving the objectives required by the legislation.

British legislation has tended to give the employer and/or the head contractor considerable responsibility, and to rely upon him to use his discretion in exercising his expertise when selecting the personnel and setting up the operational standards to achieve the objectives of the legislation.

Norwegian legislation and subordinate Regulations tend to spell out in considerable detail what manning and qualification standards the licensee should adopt. Indeed, a considerable governmental effort was given in the late 1970s to identifying manning requirements and establishing the training standards needed to ensure that those who worked offshore would be suitably trained and qualified. This task was undertaken by the Ministry of Church and Education and was seen firmly in the context of state provision of training courses.

The Leiro Committee was set up with terms of reference which required it to:[2]

1. Study qualification requirements on the basis of existing organisational plans.
2. Determine the groups of employees in the different fixed installations for whom no educational facilities were available.
3. Propose educational plans for the different groups under 2
4. Submit an estimate of costs and proposals for financing the education proposed under 3.

In the event, the new legislative framework may initiate a move towards the British philosophy of looking to the industry to ensure that personnel operate to safe systems,[3] although at present the notion of compliance with statutory minima is still very relevant to internal control.

The issues of manning, qualification and training are so interrelated, with particular details so related to the framework within which they are situated, that the general patterns of the British and Norwegian systems are best contrasted in their totality rather than attempting to deal with issues separately. However, there are within both systems two matters which lend themselves to separate treatment: namely, diving and emergency procedures. As these matters have both been dealt with in other contexts,[4] we will mention them only briefly in this chapter.

British legislation

The traditional British system of regulating for health and safety at the workplace adopted a 'checklist' approach and identified particular matters to which regulatory standards should be applied. The Health and Safety at Work, etc Act 1974 was a radical departure from this approach in that it introduced the concept of safe people in place of the earlier emphasis on safe workplaces. Sections 2 and 3 of the Act impose broad general duties on the employer. Section 2(1) requires him to ensure, so far as is reasonably practicable, the health, safety and welfare of his employees. Section 3(1) requires him to conduct his undertaking in such a way as to ensure, so far as is reasonably practicable, that persons not in his employment who may be affected thereby are not thereby exposed to risks to their health or safety.

The importance of this duty in situations where two or more organisations are working in close proximity, and the interpretation put on the relationships between Ss2 and 3 in the Swan Hunter case, have already been explained.[5] The main significance of that case for present purposes is that S2(2)(c) provides that the general duty in S2(1) extends to:

> ... the provision of such information, instruction, training and supervision as is necessary to ensure, so far as is reasonably practicable, the health and safety at work of his employees.

The Swan Hunter case established that the employer may need to *inform, instruct and train* not only his own men, but also those of a sub-contractor in order to achieve the health and safety of his own employees. Moreover, the judicial analysis in that case suggests that earlier compensation cases may give valuable guidance as to the meaning of the italicised words above. If this is so, it may be concluded that the employer must inform the employee of any dangers to which he is exposed,[6] that instruction means both telling and training,[7] and that the duty to instruct may be related to, and thus widen, other Regulations which require the employer to provide safe plant and equipment and protective clothing.[8]

It may be noted in passing that this Act also imposes, in S7, a duty on every employee while at work:

(a) to take reasonable care for the health and safety of himself and of other persons who may be affected by his acts or omissions at work; and
(b) as regards any duty or requirement imposed on his employer or any other person by or under any of the relevant statutory provisions, to co-operate with him so far as is necessary to enable that duty or requirement to be performed or complied with.

This provision is, however, of little direct comfort to the employer, or head contractor, whose workforce fail to comply with information, instruction and training given for their own safety and for the safety of others, since the criminal penalty which is the sanction for breach of this duty can only be inflicted on the worker as the outcome of proceedings instituted by the inspectorate. However, should the employee be in breach of his duty as here defined there will be little doubt that he will also have committed misconduct and he ought properly to be submitted to disciplinary procedures,[9] and the employer who fails to ensure the safe conduct of individual workers may well be in breach of his own general duties. Instruction and training should therefore stress the responsibilities of individual employees, both for the enlightenment of the employee himself, and for the protection of the employer.

It may also be worth commenting that there could be situations in which managers could be personally liable under S7 if they have failed to discharge responsibilities with which they have been properly charged; that is to say if it was appropriate so to charge them taking into account their qualifications, experience and the resources under their control.

The extension of the provisions of the Health and Safety at Work, etc Act to petroleum activities[10] arguably makes superfluous any further and more specific statutory duties concerning manning, qualifications and training. However, these general provisions operate in conjunction with the earlier and more detailed provisions of the Offshore Installations (Operational Safety, Health and Welfare) Regulations 1975 and the Offshore Installations (Emergency Procedures) Regulations 1976, which were made under the Mineral Workings (Offshore Installations) Act for the particular purpose of achieving safe systems in the offshore petroleum industry.

It is also significant that, even under the new framework legislation, it is perceived that there may be situations where more specific Regulations are necessary, to lay down systems for particularly hazardous operations; important in this context are the Diving Operations at Work Regulations 1981 made under the Health and Safety at Work, etc Act.

The Offshore Installations (Operational Safety, Health and Welfare) Regulations 1976 impose duties on the installation manager, the owner of the installation and the concession owner and relate to the operation of the installation rather than to the employment relationship. These Regulations identify some situations in which not only must persons be appointed to perform specific functions but the persons appointed must have named qualifications. Interestingly, these qualifications are not ones devised peculiarly for the petroleum industry. Thus, Reg 19 requires that radiotelephone operators must be appointed in the following circumstances and also stipulates the qualifications which they must hold:

> There shall be present on every offshore installation at any time when it is manned at least one person fully trained to be the radiotelephone operator who is the holder of a certificate of competence with respect to the equipment provided on the installation under Regulation 18(1) above issued by the Secretary of State under section 7(1) of the Wireless Telegraphy Act 1949.[11]

The Department of Energy's Guidance Note No 1 'Qualifications for Radiotelephone Operators on Offshore Installations, names six types of certificates which are acceptable to the Department: namely MPT 104, MPT 105, MPT 106, MPT 116, K515 and K516.

The minimum qualification acceptable is that conferred by certificate MPT 105. This is issued to persons who have been examined in radiotelephony and have passed in:

1. Practical knowledge of the adjustment of radiotelephone apparatus.
2. Practical knowledge of radiotelephone operation and procedure.
3. Sending and receiving spoken messages correctly by telephone.
4. General knowledge of the Regulations applying to radiotelephone communications and particularly of that part of those Regulations relating to the safety of life.

Similarly there is, in Reg 31, a specific requirement by reference to the standards of external authorities in respect of medical personnel, although it has in fact long been recognised as a somewhat weak one. It stipulates only for a registered or enrolled nurse or even a 'first-aider'. This Regulation is to some extent strengthened by the Guidance Notes on Training of Offshore Sick-bay Attendants which were produced in 1978.[12]

However, there is no legal significance in a guidance note: it is merely indicative of the enforcement agency's analysis of what is appropriate for the circumstances to which the note relates. Not only is it not possible to institute legal proceedings for breach of a guidance note,[13] it is even possible that the standards suggested in a guidance note might not be accepted in legal proceedings if the court were persuaded by strong counter-evidence. Indeed, in the case of a guidance note, unlike an approved code of practice, there is no obligation on the court to accept its contents as evidence that the law has been broken.

In future the matter will be governed by the (proposed) Offshore Installations and Pipe-line Works (First Aid) Regulations. These Regulations will be made under both the Health and Safety at Work, etc Act 1974 and the Mineral Workings (Offshore Installations) Act 1971. The requirements will be in three parts: a general set of Regulations, an approved Code of Practice, and Guidance. The Regulations set out in broad terms that, on installations and for pipe-line works (diving is excluded) and at all phases of operations, there shall be provided:

> ... such equipment, facilities and medications and such number of suitable persons as are adequate and appropriate in the circumstances for rendering first aid to and treating in accordance with the directions of a registered medical practitioner, who may or may not be present, persons who are injured or become ill while at work ...

also to

> ... provide such number of suitable persons as is adequate and appropriate in the circumstances for giving simple advice in connection with the health of persons at work.

Further the training of such persons shall be subject to qualifications that may be approved.

The suitable persons are defined in the Code of Practice as either offshore medics or first-aiders and the Code will lay down a sliding scale for the number to be appointed. There has to be at least one offshore medic for numbers of persons regularly at work up to 201 and above. First-aiders have to be appointed on the basis of one for every 25 persons. So an installation or pipe-line barge with 201 or more persons employed will have at least one offshore medic and nine persons trained in first aid.

It is considered in the Guidance that the basis for recruitment for offshore medics should be registered general or enrolled nurses or experienced ex-service leading medical assistants or equivalents.

The training standards require the selected candidate for the post of medic to undergo further training for at least four full weeks (120 contact hours) on an approved training course, the passing of a suitable examination being a prerequisite of the award of a certificate valid for three years. Subsequently, refresher courses of two weeks (60 contact hours) will have to be undertaken.

The training for first aiders is to be at least five days (30 contact hours) and they must subsequently be successfully examined on an approved course. Again, the certificate is valid for three years only. Refresher courses are in this instance of two days' duration. Such persons as have been in post for two years preceding the coming into force of the Regulations will not be required to take a full training course, but will be required to undergo refresher training and be examined within a period of two years from the Regulations coming into force.

The equipment, facilities and medication which are required are laid down in Appendix II of the Guidance. However, where there is a full-time occupational health service under the control of a qualified health physician, it is expected that the service will make decisions on the equipment, medications, etc which are adequate and appropriate in various circumstances. However, apart from laying down basic criteria in respect of the equipment in a sick bay the proposals do not require resiting or reconstruction of the sick bay.

It is interesting to note that for the purposes of the Health and Safety at Work, etc Act 1974 persons, 'are treated as being at work whether or not they are on duty'.

Apart from these rather specific examples – where particular tasks are identified and a requirement is imposed that person(s) be appointed with specialised externally-recognised qualifications – the principal manning requirements are set out under the heading of 'Operational staff' in S30 of the Operational Safety, Health and Welfare Regulations, which provides as follows:

(1) There shall be provided on every offshore installation a sufficient number of competent persons appointed by the installation manager to be responsible for the control and safety of –
   (a) the structure of the installation;
   (b) the electrical equipment of the installation;
   (c) the mechanical equipment of the installation;
   (d) lifting appliances and lifting gear;
   (e) drilling operations;
   (f) production operations;
   (g) the handling and storage of acids, caustic alkalis, explosives, radioactive and other dangerous substances;
   (h) any other unusual or dangerous operation;

and the installation manager shall ensure that a list of all such persons on the installation is maintained on the installation at a place where it can be conveniently read by persons on the installation.

It is interesting to note that none of the requirements of these Regulations relating to radiation (ie Regs 4, 7, 30 and Schedule 2) have been revoked by the new Ionising Radiation Regulations. These newer Regulations follow the pattern of Regulations made under the Health and Safety at Work, etc Act and place emphasis on training and systems of work, together with hazard assessment and contingency plans based on that assessment.

Section 30, it may be observed, does not define the task that has to be carried out except in terms of giving it a 'job title', nor do the Regulations specify the qualification necessary for a person to have before or during the time he has responsibility for carrying out any of these tasks. However S30(2) does specify that persons appointed to carry out any operation (which must include those identified in S30(1)) must be both experienced and competent, or work under someone who is. It provides:

> (2) Every person who uses any equipment (other than domestic equipment) or who is engaged in carrying out any operation (other than an operation of a domestic nature) on or near an offshore installation shall either –
>   (i) have experience of and be competent to use that equipment or have experience of and be competent to be engaged in that operation; or
>   (ii) work under the close supervision of a person who has experience of and is competent to use that equipment or who has experience of and is competent to be engaged in that operation as the case may be.

The installation manager, the owner and the concession owner, according to circumstances, bear the responsibility of deciding who, in their judgement, is sufficiently experienced or competent to be deployed to take responsibility for a particular situation. Judges tend to give a wide discretion to management in discharging its duty to select 'competent' persons to carry out statutory safety inspections, as the following dicta indicates:

> ... a man, who on a fair assessment of the requirements of the task, the factors involved, the problems to be studied, and the degree of risk of danger implicit, can fairly, as well as reasonably, be regarded by the manager ... as competent to perform such an inspection.[14]

Should an incident or accident occur which is attributable to the fault of the person so selected, there will inevitably be something approaching a presumption that the appointment did not meet the requirements of the Regulation. The burden of proof is likely to be on he who selected the wrongdoer for the task to establish that he was sufficiently competent and experienced to undertake it.

It has been common practice for British regulatory legislation to require that tasks be undertaken by 'competent' persons but the meaning of competence is not given in such legislation.[15] It is therefore a question of fact for the court to decide in the case of dispute.

The offshore installation manager has to be considered competent in relation to the management of the installation and his appointment must be in accordance with S4 of the Mineral Workings Act.

There is a further example of this regulatory technique in the Regulations here under consideration. Regulation 6 stipulates that lifting appliances and gear must be examined by an independent person, while Reg 25(2) provides:

> Any supply of drinking water on an offshore installation shall be tested for purity by a *competent person* at intervals not exceeding three months.

The 'competent person' formula is also used in respect of the requirement contained in Reg 21 that the installation manager must appoint a helicopter landing officer. Regulation 22 sets out in some detail the matters in relation to which the landing officer has to accept responsibilities, but there are no requirements as to his qualifications and training beyond that of 'competence'. However, Department of Energy Guidance Note No 2 'Guidance on the Selection, Training and Appointment of Persons as Helicopter Landing Officers' is designed to help employer and employees develop appropriate arangements in relation to their undertakings.

These Regulations also contain a general requirement in Reg 28 prohibiting the employment on offshore installations of persons under 18 years of age.

There is no restriction in British petroleum legislation on the hours which an individual may work, but Reg 29 requires an accurate record to be maintained of the hours worked by every person on the installation. Thus, while the operator in the British sector of the continental shelf need not concern himself with the careful calculation of how many persons he needs to employ to cover the man-hours he envisages will be needed, he ought nevertheless to bear in mind the possible consequences of a record of over-long working hours. It would be damaging evidence against him should the installation be endangered or damaged or persons working on it suffer personal injury, in circumstances where the records showed that abnormally long hours had been worked by individuals involved in, or responsible for, causing the accident or incident. It is worth repeating that both Regs 28 and 29, and indeed the other Regulations contained in this particular regulatory code, relate to the system of operation of the installation; while they are imposed primarily on the installation manager, in effect they require him to accept this degree of responsibility for whomsoever may be working on the installation regardless of with whom that person's contract of employment has been made.

In passing it may be noted that the Offshore Installations (Emergency Procedures) Regulations require the establishment and the maintenance by regular practice, of procedures which may be brought into operation in the event of an emergency. The essence of these procedures is that individuals be allocated particular duties and that they be trained to undertake them.

Similarly, the Diving Operations at Work Regulations 1981 identify the personnel required in the operation of a diving operation and impose specific duties on them. In particular, it should be noted that the diver himself must have a valid certificate of training issued either by the Health and Safety Executive themselves or by a body approved by the Health and Safety Executive.

All the British legislative requirements set out so far, being specific Regulations made under either the Mineral Workings Act or the Health and Safety at Work, etc Act for application to the offshore petroleum industry, apply to all installations (or activities in relation to such installations) operating in the British sector, whether fixed or mobile, and whether they are registered as ships or not. It has, however, been noticed that in some instances the qualifications adopted are those issued by another body for a more general purpose.

It is not the practice in the British sector for governmental authorities to offer training courses for the industry. Indeed, apart from the Well Control Regulations which stipulate that drilling supervisors and drillers must be certified as fully qualified concerning well pressure control, and the new first-aid training requirements when in force, there are at present apparently no training requirements drawn up by regulation exclusively for the petroleum industry. However, the Department of Energy's Guidance Notes do lay down standards of training. They do not, however, provide syllabuses – they only state what in the Department's view syllabuses should contain.

Much of the training for the British petroleum industry has been provided by the Offshore Petroleum Industry Training Board: this body has offshore training as a statutory responsibility and it has laid down standards of training for survival and general fire-fighting.

There are a number of training associations, the two main ones being the Scottish Offshore Training Association (SCOTA) and the Petroleum Training Association North Sea (PETANS). Most courses put on by the training board are SCOTA or PETANS courses. These bodies, the United Kingdom Offshore Operators Association and the similar association for divers have been active in identifying the standards of training necessary for personnel employed in the industry.

It may well be that there could be situations which have not been identified in the legislation, where a person is engaged in an activity offshore, such as operating machinery, where there is a professional qualification generally adopted onshore (whether by reference to a regulatory standard or not). In these instances, the general duties under the Health and Safety at Work, etc Act, which require that reasonably practicable standards should be adopted on almost every occasion, would tend to raise a presumption that the accepted onshore standards should be observed with respect to the employment offshore.

When an installation is registered as a British ship, the requirements of the British Merchant Shipping Acts will apply in respect of the manning, qualifications and other relevant operational matters in respect of the vessel and its crew in whatsoever part of the world it is operating.

Additionally, the Aliens Restriction (Amendment) Act 1919 stipulates in S5:

(1) No alien shall act as master, chief officer, or chief engineer of a British merchant ship registered in the UK except ... boat employed habitually in voyages between ports outside the UK;
(3) No alien shall be employed in any capacity on board a British ship registered in the UK unless he has produced to the officer before whom he is engaged satisfactory proof of his nationality.

There is no general nationality requirement imposed by British legislation in respect of employment on the British continental shelf, but it might well be a term of the licence that employment of British personnel and purchase of British goods be given preference by the licensee.

Norwegian legislation

Consideration of the Norwegian legislative requirements in relation to manning, qualifications and training in the offshore petroleum industry must in this, as in most other matters concerned with the regulation of petroleum activities since 1985, begin with the principal Act of 22 March 1985 and the Royal Decrees which immediately implemented this framework. It must then consider the individual employment rights contained in the Act of 4 February 1977 concerning worker protection and the working environment (as applied offshore by the Royal Decree of 1 June 1979), and the particular regulatory standards which remain in force after, though promulgated before, the principal Act came into force. Finally, there are special Regulations which apply only to mobile installations, some of which may be applicable only to Norwegian-registered vessels.

The principal Act (and indeed most of the Regulations relating to petroleum activities) imposes its obligations upon the licensee. Section 48 of the principal Act requires the licensee to have an organisation which is capable of managing the petroleum activities in which he is engaged.

Section 49 refers more specifically to the selection and training of personnel for employment offshore. It provides:

> The licensee shall ensure that all persons engaged in the activities possess the necessary qualifications to perform in a prudent manner the work they have been assigned. Training shall be given to the extent necessary.

Section 54 requires the licensee to give preference to the use of Norwegian goods and services. This provision is, however, intended to secure that Norwegian industry is accorded equal treatment with foreign suppliers. The use of shipping services are exempted from S54 by S38 of the Regulations of 14 June 1985.

As a quite distinct issue, a Royal Decree of 25 August 1967 on control of foreigners remains in force and imposes stringent requirements concerning the granting of work permits to other than Norwegian citizens to work on the Norwegian sector of the continental shelf.[16] This immigration control measure does not, in practice, prevent the employment of 'foreign experts'.

Commitment to the development of the Norwegian economy and the reservation of employment opportunities for Norwegian citizens is apparently strengthened by S37 of the Regulations of 14 June 1985 implementing the principal Act. This section requires that the Norwegian language shall be used to the greatest extent possible in petroleum activities. However, this is justified in the interests of safety. Section 40 spells out to a greater extent the requirement to use Norwegian labour:

> With the exception of the licensees' contractors' and sub-contractors' permanent workforce and the minimum number of specialists required, labour shall be hired through the public labour exchange in Norway.
>
> . . .
>
> The licensee is responsible for ensuring that all foreign employees who are required to have a work permit possess the necessary permit before work is commenced.

The section goes on to impose on the licensee a considerable responsibility for training the offshore workforce, in the interests of improving the qualifications of the national labour force:

> When the total number of employees in the operator's organisation in Norway exceeds 20, the licensee shall present for approval to the Ministry of Local Government and Labour a plan for the training and qualification of the employees. The objective shall be, *inter alia*, to raise the level of competence among Norwegian

employees so that they may take over positions at all levels in the operator's organisation. The licensee shall also ensure that contractors and sub-contractors performing major work for him present similar plans.

It is clear that 'in Norway' as used in the above means both on the mainland as well as petroleum activities on the continental shelf. The system of internal control (below) imposes similar obligations on the licensee; and the licensee's duty to train is itself repeated at greater length, but with no greater specificity, in S24 of the Regulations concerning safety made by Royal Decree on 28 June 1985:

> The licensee shall ensure that all personnel involved in the petroleum activity have the necessary qualifications for proper performance of work they have been assigned. Training shall take place to the extent necessary and under satisfactory supervision.
>
> For positions with significant importance to safety, job qualifications shall be described.
>
> The licensee shall ensure that all present on installations or vessels participating in the activity have adequate training and practice in connection with emergency preparedness situations.
>
> The Ministry may stipulate requirements regarding qualifications through regulations or specific administrative decisions.

However, the following section (S25) requires the licensee to prepare an organisational plan for the relevant phase of the activity for which he is seeking consent (ie exploration, construction, production, removal). This plan must describe lines of responsibility, authority and communication, functions and work areas.

Finally, in this framework legislation, S7 of the Regulations concerning the licensee's internal control contained in a Royal Decree of 28 June 1985 requires the licensee's internal control to ensure, *inter alia*:

> ... that competent personnel is used during planning and implementation of the separate phases, including the design, fabrication, installation and operation;
>
> that the licensee's and contractors' employees are given necessary training;
>
> that information and documentation shall be presented to the authorities on time in accordance with the laws, regulations and guidelines;
>
> that lines of responsibility, authority and communications are clearly defined and understood.

Just as the Health and Safety at Work, etc Act is arguably a sufficient framework for ensuring the adequacy of the manning, qualification and training to achieve safe systems of work in the offshore petroleum industry in the British sector, so too this framework, taken in conjunction with the extension offshore of the Act of 4 February 1977 on worker protection and the working environment, is arguably a sufficient legislative provision for the Norwegian sector. However, the present enforcement policy is still wedded to the earlier regulatory standards.

The Working Environment Act of 4 February 1977 makes provision for the contractual arrangements of the individual worker. Both S7 of the Reg on the licensee's internal control and S4 of the Royal Decree of 1 June 1979 concerning the extension of the principal Act to the offshore industry, impose duties on the licensee to ensure that the relevant provisions of that Act are observed (S7) and that workers are informed of their rights under that legislation (S4).

The legislation is particularly important in the present context because of the stringent provisions it contains concerning working hours. Not only will the Act impliedly affect the contracts of individual workers, but it must clearly have considerable implications for the licensee in planning his manning requirements; if his workforce can work only restricted hours he must maintain a sufficient number of workers to enable a system of work within these restrictions.[17]

In the current Norwegian shelf legislation, the following matters concerning the manning, qualifications and training of personnel are of importance: medical provisions; Regulations for crane operations; Regulations for drilling; Regulations for diving; Regulations for life-saving appliances; Regulation for aeromobile radio equipment; Regulations for helicopter decks. Each of these matters will be reviewed here but only to the extent that they have not been explained at greater length in other chapters.

Medical provisions[18]

The regulatory provisions concerning the protection of the health of those who are employed offshore are of two sorts: first, those which are intended to ensure that only those who are physically fit are engaged to work offshore; and second, those which concern the employment of medically qualified personnel offshore.

Regulations concerning medical examination of employees stipulated on 1 August 1980, require those who are to work offshore to undergo medical examination. Provisional standard directives for state-registered nurses employed on installation for production, etc stipulated on 11 April 1978, and standard instructions for physicians in charge of fixed installations stipulated in May 1982, make provision for suitably qualified medical personnel to be employed offshore.

Crane operations

Regulations for cranes on production platforms stipulated 25 May 1977, and similar regulations for deck cranes on Norwegian and foreign drilling units of 31 January 1978, stipulate that cranes shall be operated only by personnel having an approved crane operator's licence. To obtain a licence a person must be at least 20 years of age and have passed a special medical examination. He will have to have at least 150 logged hours in offshore crane operation as a trainee before he will be granted a licence.

There are also provisional guidelines of 23 November 1978 for the training and furnishing of certificates of competence for crane operators on mobile drilling units and other mobile installations.

Drilling

Regulations for drilling which apply to both fixed and mobile installations of 23 September 1981, require the licensee to present an operational plan prior to commencement of either production or exploration drilling. It is required that this plan must:

> ... clearly state the command system and the line of command during normal operations as well as during emergencies. The plan should be incorporated into the organisational plan for the other operations on the installation.

It is also required that:

> During operations on board installations ... a person in charge shall always be present on the installation. The organisation plan shall indicate the qualifications of the person in charge as well as of his immediate subordinates.

Regulations of 22 February 1983 state the qualification requirements of drilling personnel. These Regulations identify the various jobs which will be undertaken by personnel in a drilling team and specify the qualification requirements for those employed in these tasks. For example, a floorman/roughneck (a person who performs work on the drill floor, instructed by the driller and the toolpusher) either must have had six months' experience on production/drilling platforms as a roustabout or in a similar capacity, and have completed a course for floormen at a recognised educational institution (or 'six weeks' Drilling Course I'), or must have had 10 weeks' experience on production/drilling platforms as a roustabout or in a similar capacity, and have undertaken the first year of technical school on a drilling technology course.

Senior personnel, including drilling supervisor and toolpusher, must also have undertaken appropriate training to make them fully qualified concerning pressure control. They must take an annual examination in pressure control at a recognised educational establishment. In fact, the requirements are very similar to those in the British Regulations.

Ultimately it is the licensee's responsibility to ensure that all the personnel in the drilling crew are qualified in accordance with the Regulations. These Regulations must be seen in conjunction with the Maritime Directorate's Regulations for qualifications and manning of the ship's crew on Norwegian drilling and other mobile units (below).

Diving[19]

There appear to be two sets of Regulations for diving currently in force: the provisional Regulations of the Petroleum Directorate stipulated on 1 July 1978 and the Regulations of 10 April 1984 from the Maritime Directorate for mobile installations. The latter are part of shipping laws rather than shelf laws, however, and are not confined in their application to the Norwegian continental shelf nor are they confined to diving in connection with petroleum activities. Both sets of Regulations lay stress on safe systems of operation, but the Petroleum Directorate's Regulations, like the British Diving at Work Regulations are largely concerned with personal duties and matters of personal competence.

There are particular restrictions on the working hours of divers: these are contained in the Regulations concerning working hours for diving personnel on drilling platforms of 26 June 1981.

Live-saving procedures[20]

Regulations of 3 February 1982 made for life-saving appliances on fixed installations make provision for manning, musters and drills, providing particularly that every emergency station must be under the command of a qualified person, designated in advance. Similar provisions for mobile installations are contained in the Regulations of 10 September 1973.

Aeromobile equipment

Regulations issued by the Norwegian Telecommunications Administration on 8 September 1980 require the radio station on drilling and accommodation platforms to be operated by a radio operator, and prescribe his responsibilities. The radio operator is required to have a radiotelegraph operator's certificate of at least second class. If the installation is self-propelled, the radio operator must have a certificate in accordance with Regulations in force for the merchant marine.

Similar Regulations of 10 September 1980 apply to production platforms.

Helicopter landing officer

Regulations for helicopter decks on drilling platforms made on 18 April 1973 require, in similar terms to the equivalent British legislation, that a person must be appointed to take responsibility for operations conducted on the helicopter deck and set out his duties.

In addition to the above there are also two sets of detailed Regulations, one on manning and one on qualifications, laid down on 23 March 1982 by the Maritime Directorate for mobile installations. The Maritime Directorate has also, in Regulations of 11 December 1981 provided standards for certificates of competency for the ship's crew on drilling vessels and other floating installations. These Regulations were described as applying to Norwegian and other mobile installations. In so far as they apply to Norwegian installations, they will be regarded as a part of flag law and applied on a world-wide basis. It is not clear to what extent they remain directly applicable to foreign mobile installations, since the operational arrangements of mobile installations on the Norwegian continental shelf are now a matter for the Petroleum Directorate. It seems likely that the standards in these Regulations would at least be guidelines to which the Petroleum Directorate would turn when evaluating the documentation presented to it by a foreign mobile installation wishing to operate in the Norwegian sector.

The manning Regulations state the number and composition of the crew and set out the responsibilities and line of authority for the platform manager, the drilling section, the machinery and maintenance section, the hull and stability section, the catering and services section, the radio officer, the safety officer and the medical warden. The Regulations on qualification requirements deal with the necessary qualifications for largely the same named crew members, setting out the necessary training and qualifications in considerable detail. As might be expected, there are many similarities between the qualification standards stated in these Regulations and those contained in the shelf standards which have already been considered.

References

1   See Chapter 9.
2   See Reports of the Leiro Committee, set up under the chairmanship of G Leiro (Chief Education Officer, Rogoland) 3 August 1977.
3   There are basic offshore training courses provided by institutions such as the Rogoland Maritime School (Stavanger) which provide the knowledge of offshore conditions which is expected to be held by anyone seeking employment offshore in the Norwegian sector – even those seeking 'domestic employment'. Currently the Norwegian Industry Association for Oil companies (NIFO) has evaluated and made proposals for changes in the basic training course. See Norwegian Petroleum Directorate's *Annual Report* for 1985 at p81.
4   For diving see Chapter 8 and for emergency procedures see Chapter 13.
5   See Chapter 7.
6   *Stokes v GKN Ltd* [1968] 1 WLR 1776.
7   *Boyle v Kodak* [1969] 2 All ER 439.
8   *Bux v Slough Metals Ltd* [1974] 1 All ER 262.
9   See Chapter 10.
10 The Health and Safety at Work, etc Act 1974 (Application outside Great Britain) Order 1977.
11 Reg. 18(1) refers to signalling equipment.
12 These standards are comparable to those applying onshore in factories at the time. Unfortunately it has since been recognised that this 'first-aiders' training as under 2(a)(ii) has limitations, even in onshore applications.
13 Proceedings would be for breach of a legal duty, eg the general duty in S2(1) of the Health and Safety at Work, etc Act.
14 *Brazier v Skipton Rock Co Ltd* [1962] 1 All ER 955 per Winn J at p957.
15 See Chapter 7.
16 Canadian law has adopted a similar approach and this was a factor in the Ocean Ranger disaster. The protection of jobs for nationals of the coastal state can be dangerous if the relevant skills are not readily available among the local population.
17 See Chapter 12 for detailed description of hours legislation.
18 See also Chapter 11.
19 See Chapters 7 and 8.
20 See Chapter 13.

11

Safe Systems: Health Provision

Principal legislative provisions

British legislation
Mineral Workings (Offshore Installations) Act 1971.
Health and Safety at Work, etc Act 1974.
Public Health Act 1936.
Misuse of Drugs Act 1971.
The Offshore Installations (Construction and Survey) Regulations 1974.
The Offshore Installations (Operational Safety, Health and Welfare) Regulations 1976.
The Offshore Installations (Emergency Procedures) Regulations 1976.
The Misuse of Drugs Act 1971 (Modification) Order 1985.
The Misuse of Drugs Regulations 1985.
The (Proposed) Offshore Installations and Pipe-line Works (First Aid) Regulations 1987.

Norwegian legislation
Act of 4 February 1977 relating to worker protection and working environment.
Regulations concerning safety, etc made by Royal Decree of 28th June 1985.
Royal Decree of 25 November 1977 relating to hygiene, medical equipment and medicines.
Regulations concerning labelling, etc of chemical substances and products which may involve a hazard to health made by Royal Decree dated 25 November 1982.
Regulations concerning medical examination of employees of 1 August 1980.
Supplementary Regulations covering requirement concerning potable water systems of 23 October 1978.
Provisional standard directives for state-registered nurse employed on installations for production, etc of 11 May 1978.
Standard instruction for physician in charge of fixed installations on the Norwegian continental shelf of May 1982.
Provisional Regulations for living quarters on production installations of 2 April 1979.
Regulations concerning the construction and equipment of living quarters on drilling units and other mobile offshore installations of 11 June 1982.

Regulations on arrangements on and below deck on drilling units of 31 January 1978.
Regulations concerning potable water systems on mobile drilling units of 7 January 1985.
Provisional standard directives for state registered nurse on mobile drilling platforms and
other mobile offshore installations of 11 May 1978.
Regulations providing for medical examination of persons employed in Norwegian-
registered ships of 30 March, 1981.

The nature of the problem

Employment of persons offshore demands special consideration of special health
issues. Normal working conditions offshore involve a certain amount of stress
of a nature which is not usually found in onshore employment: having persons
living together for spells of offshore duty has the potential for creating a
stressful environment. Moreover accommodating numbers of persons in close
proximity for at least a week, and more often a fortnight at a time creates an
environment where epidemics can readily occur.

In addition to this there are particular hazards related to the work, some of
which are similar to those found in heavy engineering onshore, others related to
the nature of petroleum are more akin to those found in the onshore chemical
industry.

The isolation of the workplace from regular medical facilities necessitates
fuller provision of first-aid facilities than would normally be required for a
workforce of a similar size in, for example, a factory onshore. Removal from the
installation to a hospital may not be immediately possible: in the best
circumstances the journey from the installation to the hospital is likely to take
some hours.

Finally, catastrophe offshore involving evacuation of the installation, must be
provided for, with adequate emergency provisions to care for the injured and
otherwise sustain the evacuees.

In many respects these problems are very like those experienced in
employment in the merchant shipping industry, but to the problems of the
shipping industry must be added those of onshore heavy engineering.

The emphasis of the regulatory system must be on maintenance of good
health and prevention of accidents. Much of the preceding chapters in this book
has been concerned with safe systems for the prevention of accidents. In the
event of an accident, or the occurrence of illness, emergency procedures may
need to be called into effect, and these will be considered in the final chapter. This
chapter will consider mainly that part of the regulatory system which is
concerned with the maintenance of a healthy workforce.

Framework of legal control

British legislation

The Health and Safety at Work, etc Act S2(1) imposes on an employer the general duty to do what is reasonably practicable to ensure the health, safety and welfare of his employees, and this duty is applicable offshore. Particularly relevant is S2(2)(e) which provides:

> S2(2) Without prejudice to the generality of an employer's duty under the preceding sub-section, the matters to which that duty extends include in particular –
> (e) the provision and maintenance of a working environment for his employees that is, so far as is reasonably practicable, without risks to health, and adequate as regards facilities and arrangements for their welfare at work.

Section 3 of this Act imposes on the employer a duty to conduct his undertaking in such a way as to ensure, so far as is reasonably practicable, that persons not in his employment who may be affected thereby are not thereby exposed to risks to their health or safety.

Strictly speaking, these two sections are sufficient to impose upon the operating company the duty not only to protect those in its direct employment but also to protect those who are employed by sub-contracting organisations, particularly now that it has been judicially confirmed that the head contractor's duty requires him to instruct and inform other contractors' employees to the extent necessary to ensure their health and safety.[1]

Should current proposals to make a general regulatory provision for employers onshore to assess their workplaces in order to identify any substances hazardous to health, and to control any hazards which the assessment reveals, be brought into force and extended offshore, this Act will impose very wide obligations upon offshore operators.[2]

Moreover, it is arguable that some onshore Regulations, such as the Classification, Packaging and Labelling of Dangerous Substances regulations 1984, although not actually extended offshore, might be used in evidence as illustrative of what is 'reasonably practicable'.[3] Nevertheless there is a number of specific regulatory provisions in Regulations made under the Mineral Workings (Offshore Installations) Act which must also be considered.

Norwegian legislation

In Norway the Worker Protection and Working Environment Act is largely applicable to fixed installations, but appears to be less relevant in respect of health issues than some of the specific offshore regulatory provisions. However, it is interesting to note that the general Regulations concerning labelling, etc of chemical substances and products which may involve a hazard to health, which were made under this Act by Royal Decree of 26 November 1982, are deemed to have offshore application and are thus specifically incorporated by the Petroleum Directorate in their manual of legislation applicable to petroleum activities.

Section 23 of the Regulations concerning safety which were made by Royal Decree on 28 June 1985 provides:

The Ministry of Social Affairs may establish further provisions concerning health aspects, including hygiene, the supply of potable water, the preparation and availability of nourishment, hygiene standards, and other factors of significance for health and hygiene.

The Ministry of Social Affairs may stipulate further provisions concerning preparedness relating to health care and health services, including sick bays and medical offices, medical equipment and transportation of the sick and injured. Further provisions may also be stipulated concerning requirements relating to workers' health and to qualifications and training of personnel to ensure that specified aspects are attended to ...

This section is largely declaratory of the powers which had previously been vested in the Ministry of Social Affairs and, until such time as the older regulatory provisions may be updated, those which were made before the Royal Decree of 1985 will remain in force. Especially important are the Regulations relating to hygiene, medical equipment and medicine, etc on production installations made by Royal Decree of 25 November 1977.

Implementation of this framework

The implementation of the framework requires that safe systems be established to ensure:

1. The selection of persons suitable for offshore employment.
2. A healthy environment on the installations
   - good and wholesome food and clean drinking water
   - clean and adequate living and accommodation standards.
3. A suitably equipped and staffed sick bay.
4. Provision for evacuation of those who cannot be treated offshore.

Selection of personnel

Careful selection of personnel to ensure the employment of persons who are physically and temperamentally suited to employment offshore is a valuable precautionary measure to reduce the incidence of work-related ill health, particularly as the potential for hazards created by drug and alcohol abuse is an especially serious consideration.

British legislation

In the British regulatory system it is not usual for pre-employment medicals to be required by law. If they are required it is because the prospective employer makes his offer of a contract of employment dependent on a satisfactory medical report. The British offshore petroleum industry is not subject to any special regulatory requirements in this respect, but the nature of employment offshore would appear to make it desirable from the employer's point of view to require both a pre-engagement medical and periodical medical examinations during the performance of the contract. Indeed, in order to detect drug abuse in particular, searching of persons and the property of employees may be considered desirable: if this is the case the employer ought to ensure that he is expressly given these powers under the contract, since it is not clear that the employer would otherwise have the right to carry out such searches.

Norwegian legislation
Under the Norwegian regulatory system a general requirement for pre-employment medical examinations was stipulated in the Royal Decree of 25 November 1977. Subsequently, detailed requirements were set out in Regulations issued by the Ministry of Social Affairs on 1 August 1980. Additionally, the Royal Decree on safety of 28 June 1985 grants powers to the Ministry of Social Affairs to make Regulations concerning a wide variety of health matters offshore.

Regulations so far made require any person who is to be employed to work offshore on a production platform – other than those who are casually employed for tasks of less than three days' duration – to undergo a prior medical examination and to have a certificate to prove that he has passed the test. The examination has to be carried out in accordance with the guidelines issued by the Directorate of Health and the results of the examination have to be entered on the special form provided by the Directorate. Only authorised doctors are empowered to carry out the examination, which must cover sight and hearing and, more generally, the doctor must confirm that the person concerned is 'not suffering from any disorder which renders him/her unfit for the work involved or which may endanger the health and safety of other employees'. The person undergoing examination has to make a personal statement as to his health.

The Regulations contain a schedule of medical conditions which may render an employee unfit for service offshore including, for example, obesity, diabetes and leukaemia. Some of the conditions listed are to be regarded as complete bars to offshore employment on a production platform; other conditions are not an absolute bar and the employee may be given a restricted health certificate authorising employment only in restricted areas of work.

A health certificate is only valid for a limited period of time. When initially seeking employment the employee's certificate must not be more than six months old; if the employee is over 50 the period of validity is only three months. In the event of continuous employment the certificate must be renewed every two years; where the employee is over 50 the certificate must be renewed annually. If an employee has been incapable of work for six weeks because of illness or injury a new certificate is almost invariably required.

The employer is under a duty to ensure that the employee has a valid certificate and that such medical examinations are carried out as are necessary to renew the certificate. The licensee must keep the health certificate during the contract of employment: the Health and Petroleum Directorates are both entitled to ask to see the certificate to verify that it is valid.

Similar Regulations issued by the Directorate of Seamen on 30 March 1981 provide for medical examination of persons to be employed on Norwegian-registered ships. These apply to mobile installations which carry the Norwegian flag. Section 26 of the Seamen's Act also requires medical examinations for sailors.

There appears to be no official requirement for medical examination of persons employed on foreign mobile installations operating in the Norwegian sector. However, it would apparently be within the power of the Petroleum Directorate to require such medical evidence of both the ship's and the drilling crews before granting approval to a foreign mobile unit to operate in the Norwegian sector.

A healthy environment: food and drinking water
British legislation

The provisions in the British regulatory system are expressed in very general terms. The general requirements concerning equipment in Schedule 2 Part VIII of the Construction and Survey Regulations[4] are deemed sufficient to ensure that there is adequate and satisfactory plant and equipment for the provision of a good water supply and for the storage and preparation of food. That this plant and equipment is utilised properly when the installation is in operation is intended to be secured by the provisions of the Offshore Installations (Operational Safety, Health and Welfare) Regulations 1976 Part III, Health, where it is stated:

Provisions
26 All provisions for consumption by persons on an offshore installation shall be fit for human consumption, palatable and of good quality.

The same British Regulations make general provision for drinking water:

Drinking water
25(1) There shall be provided and maintained on every offshore installation, at suitable points clearly marked 'Drinking Water' conveniently accessible to all persons on the installation, an adequate supply of clean wholesome drinking water.
(2) Any supply of drinking water on an offshore installation shall be tested for purity by a competent person at intervals not exceeding three months.

Formerly, provision could be made for food hygiene in the instructions to be given to the licensee under the model clauses set out in the schedules to the Petroleum (Production) Regulations 1976 requiring the licensee to meet the onshore standards of the Public Health Act 1936. The Food and Drugs Act was similarly used as a benchmark for the licensee in the past, but the onshore Act of 1955 has been repealed, the Petroleum Regulations are now dated 1982, and no provision has been made for employing the 1984 Food Act under the Regulations: indeed it is not established that the earlier legislation was actually so employed in licences. Although it is possible that there may be early licences still in use which contain these provisions, the matter is for present purposes governed by the Operational Safety, Health and Welfare Regulations, Reg 7 of which (which enables written instructions to be given) must be read in conjunction with Schedule 2, where the matter is dealt with specifically.

Norwegian legislation

In contrast, Norwegian Regulations make detailed provisions for water standards on both fixed platforms and mobile installations. The primary source of these Regulations for production installations is the Royal Decree of 25 November 1977. This, in S5, makes the general requirement concerning potable water that an installation shall have an adequate and hygienically satisfactory supply and requires its maintenance and testing, including the requirement that sewage and effluent shall be discharged in such a manner that they do not contaminate the intake of seawater from which potable water is produced. Supplementary Regulations issued by the Ministry of Social Affairs on 23 October 1978 contain more detailed requirements concerning potable water systems and provide guidelines for disinfection of potable water. Potable water is defined in these Regulations as water intended for drinking and the preparation of food, as well as water used in sanitary installations, but not including water for flushing toilets and urinals.

The Regulations require that the potable water supply, including intake, treatment plant, methods of disinfectant, and transport systems shall be approved by the Directorate of Health. They require that the system should have sufficient capacity to guarantee a supply of at least 250 litres per person per day for personnel who live and work on the installation and 150 litres per person per day for workers and visitors who do not live thereon.

They require the materials used in construction of pipes and tanks to be of such a nature that they do not release substances into the drinking water in such quantities that health is endangered. The detailed Regulations for the construction, use and maintenance of the system include requirements that the water tanks must be kept clean, must have ready means of access for inspection and must be drainable. Section 18 also requires that the potable water system must be arranged so that infection or other forms of pollution of treated drinking water cannot occur. Section 22 requires that disinfection of the water shall be checked once every day, or more frequently if necessary.

Water taken on board from ashore via a supply ship must be disinfected by chlorination. Water may also be disinfected by ultraviolet radiation, provided that the equipment used is approved by the Directorate of Health. Silver ions may be used for disinfection provided that the process set out in the Regulations is followed.

Similarly detailed requirements for mobile installations were laid down by the Maritime Directorate on 7 January 1985.

The standards to be met in respect of dining rooms and kitchens are also spelt out in considerable detail: requirements are set out for production installations in the Royal Decree of 25 November 1977. Section 6 makes detailed provisions for kitchens and food stores. The section provides both for their siting in relation to other rooms and for their cleanliness, requiring for example, that:

> Floors, walls, and ceilings shall have unbroken surfaces and they, as well as the tables, shelves and other furniture shall be made from materials which are easy to keep clean.

Stipulations are also made about toilet and handwashing facilities for kitchen staff and also for dishwashing arrangements.

Section 8 requires that refuse and other household waste shall be collected and kept in a sufficient number of closed containers which shall be easy to keep clean, until such waste is finally disposed of in accordance with the requirements of the Ministry of Environment.

Section 9 requires high standards of personal hygiene from personnel who are handling food. They are required to wear suitable protective clothing which is clean and in good repair and which is not used outside the kitchen area.

A healthy environment: living and accommodation standards

Both the British and the Norwegian regulatory systems have provisions concerning the standard of accommodation on installations, one objective of which is to protect the health of those who live and work on them.

British legislation

The British Offshore Installations (Construction and Survey) Regulations 1974, Schedule 1 Part V, paragraph 6 provides:

Every offshore installation shall be provided with accommodation –

(a) placed and constructed so as to afford persons on the installation protection from weather, fire, noise and vibration;
(b) sufficient in area to meet the needs of the maximum number of persons likely to be on board the installation at any one time; and
(c) containing facilities and equipment for that number of persons as respects sleeping, food and water storage, food preparation and consumption, sanitary and recreational requirements.

Guidance provided in the *Blue Book* provides for more specific standards and these are followed by the certifying authorities: they are similar to those set out in detail in respect of the Norwegian law (below).

Norwegian legislation

The Norwegian provisional Regulations for living quarters on production installations which were issued by the Petroleum Directorate on 2 April 1979 not only contain standards relating to heating, ventilation and noise but specify detailed standards for bedrooms, dining rooms, recreation rooms, changing, shower rooms, lavatories, refrigerating and freezing chambers. The bedroom requirements contained in S8 are particularly detailed. At a relatively general level it is provided in S8.1 that a bedroom shall be fitted out for a maximum of two persons; bedrooms intended for one person shall have a minimum floorspace of 6m^2 and bedrooms intended for two persons shall have a minimum floorspace of 12m^2. Section 8.3, 'Inventory and miscellaneous equipment for bedroom' includes such matters as:

Bedrooms shall be equipped with two lockable wardrobes for each bed. Each wardrobe shall have an inside height of at least 1.80m and a cross section of at least 0.60 × 0.50m.

Wardrobes shall be designed and equipped for installation of shelves and arrangement for hanging clothing under.

The Maritime Directorate, by Regulations issued on 11 June 1982 (shelf) and 13 January 1986 (flag), has laid down similarly detailed standards concerning the construction and equipment of living quarters on drilling units and other mobile offshore installations. For example, S13 provides:

Furniture and other equipment for cabins:
1 The berths shall be at least 2000mm long and 800mm wide, both measurements made internally.
2 The vertical distance between berths and between the upper berth and the ceiling shall be at least 900mm.
3 In each cabin there shall be a lockable closet for clothes. In addition there shall be a separate closet for the survival suit.
4 The closet shall have an inside height of at least 1800mm and an inside area cross-section of at least 600 × 600mm.
5 In each cabin there shall be at least two lockable drawers of 0.06cu m.
6 Cabins shall be equipped with a table and comfortable seating.
7 In cabins intended for more than one person, berths shall be furnished with curtains made of approved material...

In addition, Regulations on arrangements on and below deck on Norwegian and foreign drilling units of 31 January 1978 (shelf) and 13 January 1986 (flag) contain general provisions for the standard of living accommodation on such matters as lighting and ventilation.

The Royal Decree of 25 November 1977 S11 stipulates hygiene requirements which include provisions relating to bed linen and mattresses. Sheets, pillow-cases and quilt cases must be changed whenever a new person occupies a bed. Bed linen must be changed at once if it is soiled or dirty; in any case once a week.

Sick bay and medical staffs

Both British and Norwegian law make provision for sick bays and for personnel and systems for the treatment of those who suffer accidents or illness.

British legislation
In the British system the Offshore Installations (Operational Safety, Health and Welfare) Regulations 1976 Part III, Health, Reg 27 made what are – for the British regulatory system – detailed requirements for sick bays. However, new standards will shortly be brought into force by the proposed Offshore Installations and Pipe-line Works (First Aid) Regulations: these new standards, which will consist of general Regulations, a code of practice and guidance will also cover pipe-laying barges.[5] There are also in preparation Guidelines on occupational health services for production installations.

Norwegian legislation
Uncharacteristically, the Norwegian Regulations of 2 April 1979 on living accommodation on production platforms make reference to sick bays, but provide no details of the standards required. This matter has been delegated to the Ministry of Social Affairs. However, any particular Regulations this Ministry may have made are not published in the Petroleum Directorate's

handbook. The matter is dealt with relatively generally in the Royal Decree of 25 November 1977 Chapter III. Section 18 requires that installations must have a health office and continues:

> On major installations the health office shall consist of one or more consulting rooms/ treatment rooms, sick room(s) and work room for the nurse. The health office must be furnished and equipped in such a manner that the health personnel will be able to carry out their duties in a satisfactory manner.

> The rooms must be of adequate size, and none of them shall have a floor area smaller than 15 sq m.

> Consulting rooms shall have the following equipment:

> an examination/treatment couch, a work table for the nurse, a lockable medicine cupboard, a hand basin with fittings for mixing hot and cold water, emergency lighting and an alarm system.

> Consulting rooms shall have emergency lighting and a power supply sufficient to permit work to continue there in emergencies.

> At least one of the consulting rooms shall have a lockable cupboard for dangerous drugs, a refrigerator for proper storage of drugs which need refrigeration, a laboratory area and a permanent telephone with its own telephone number.

> Consulting rooms and work rooms shall have essential office furniture and equipment and lockable filing cabinets.

> The sick room(s) shall be furnished and equipped to permit short term in-patient care, eg while waiting for transport. Sick rooms shall not have more than two beds each.

> In connection with the sick room there must be a bathroom with shower, toilet and hand basin with fittings for mixing hot and cold water. There must also be one bath tub for the treatment of hypothermia.

> The number of beds for in-patients shall be calculated in relation to the total number of persons present at the installation at any time. The beds must be placed so as to allow free access from at least three sides. The sick room shall have a wardrobe locker, a bedside table and an emergency calling system. The emergency calling system shall be connected to a permanently staffed room and to the bedroom of the nurse on duty.

> On smaller installations the functions of the health office may be combined in one room of adequate size, at least 20 sq m.

> ...

> The health office shall be adequately insulated against noise and have heating, ventilation and lighting ...

> The health office should as far as possible have access to daylight.

> The health office shall not be used for any other purpose.

> The health office shall be situated in such a manner as to permit safe and reliable stretcher transport into and out of the rooms.

Regulations for the construction and equipment of living quarters on drilling units and other mobile offshore installations were laid down by the Norwegian

Maritime Directorate on 11 June 1982 (shelf) and 13 January 1986 (flag). These include in S15 outline provisions for sick bay accommodation: these provisions are much less detailed than those given above for fixed platforms.

On Norwegian production platforms the governing provisions are in provisional standard directives laid down by the Ministry of Social Affairs on 11 May 1978, although these directives are only guidelines and have to be read in the context of the Royal Decree of 25 November 1977, S16 of which requires that a registered nurse shall be employed on manned installations. It is interesting to note that the directives place some emphasis on preventive care as opposed to curative treatment, suggesting that the nurse:

> ... shall make her/himself familiar with health conditions at the installation, and shall conduct visiting, informational, advisory, executive and inspectoral activity for the protection of health ...

There is a similar provisional standard directive of 11 May 1978 for mobile installations.

Standard instructions given in May 1982 by the Director General of Health give a job description for a physician in charge of a fixed installation. There appears as yet to be no statutory requirement that a physician be appointed for every fixed installation but it would appear that these instructions are indicative of an official wish to upgrade medical provision offshore in the Norwegian sector.

Evacuation

Medical provision offshore is intended to be little more than first aid and persons who are incapable of duty will be sent back to shore as soon as possible. From the earliest Norwegian Royal Decrees (see now Regulations concerning safety of 28 June 1985) the licensee has been required to give particulars of the arrangements he has made for onshore medical treatment for those who suffer incapacity offshore.

The standing arrangements for assistance from shore is among the matters which should be provided in the British system as rules in the emergency procedures manual under Reg 4 of the Offshore Installations (Emergency Procedures) Regulations 1976. In whatever offshore regime an operator is working, whether or not there is any legal requirement that such arrangements be made, it is clearly prudent that the licensee should ensure that there are adequate links with onshore provision. These matters will be considered more fully in Chapter 13.

References

1 *R v Swan Hunter Shipbuilders Ltd* [1981] ICR 831.
2 A consultative document containing draft Regulations has been published on *Control of Substances Hazardous to Health*.
3 SI 1984 No 1244.
4 See Chapter 9.
5 See Chapter 10.

12

Welfare and Terms of Employment

Principal legislative provisions

British legislation
Aliens Restriction (Amendment) Act 1919.
Continental Shelf Act 1964.
Mineral Workings (Offshore Installations) Act 1971.
Health and Safety at Work, etc Act 1974.
Employment Protection Act 1975.
Employment Protection (Consolidation) Act 1978.
Employment (Continental Shelf) Act 1978.
Social Security and Housing Benefits Act 1982.
Merchant Shipping Acts 1894–1984.
Wages Act 1986.
The Employers' Liability (Compulsory Insurance) General Regulations 1971.
The Offshore Installations (Application of the Employers' Liability (Compulsory Insurance) Act 1969) Regulations 1975.
The Employers' Liability (Compulsory Insurance) (Offshore Installations) Regulations 1975.
The Employment Protection (Offshore Employment) Orders 1976-81.
The Offshore (Operational Safety, Health and Welfare) Regulations 1976.
The Employment Protection (Offshore Employment) (Amendment) Orders 1977 and 1981.
Oil and Gas (Enterprise) Act 1982.
The Statutory Sick Pay (Mariners, Airmen and Persons Abroad) Regulations 1982.
The Merchant Shipping (Safety Officials and Reporting of Accidents and Dangerous Occurrences) Regulations 1982.

Norwegian legislation
Act pertaining to petroleum activities of 22 March 1985.
Worker Protection and Working Environment Act of 4 February 1977.
Regulations applying the Seamen's Act to mobile installations of 21 October 1976.
Licensee's internal control by Royal Decree of 28 June 1985.
Regulations for safety delegates and working environment committees made by Royal Decree of 29 April 1977.

Regulations relating to worker protection and working environment made by Royal Decree of 1 June, 1979.

Regulations relating to safety manning in the event of an industrial dispute on the Norwegian continental shelf of 19 March 1982.

Regulations on working hours on board Norwegian drilling platforms and other mobile installations of 19 August 1977.

Regulations on the keeping of a journal for working hours for drilling units and other mobile offshore installations of 11 May 1978.

Regulations concerning protection supervisors and protection and environment committees on board ships of 15 November 1976.

Regulations concerning the scope of the Seamen's Act by Royal Decree of 19 December 1980.

So far, this account has been almost entirely concerned with statutory regulation of working conditions offshore. Arguably, however, statutory control is only necessary where, and because, the bargaining process on which the contract of employment is conceptually based, has not otherwise resulted in satisfactory conditions of employment.[1] This chapter will be mainly concerned with the contract of employment and it will therefore consider matters which lie on the boundary of regulatory control.

In the English legal system it is still customary to regard employment as a relationship which is based on a legally binding contract freely negotiated by the parties, even though contractual freedom has been severely restricted by protective legislation – legislation concerned not only with health and safety but also with protecting the worker's property in his job. This chapter will analyse the employment relationship from the English common lawyer's perspective of the formation, content and termination of a contractual relationship.

The main subjects for discussion will be contractual terms in the offshore context, applicable employment protection legislation (such as protection against unfair dismissal or compensation for redundancy), sick pay, restriction on hours of work and length of offshore tours of duty, and collective employment law, such as the right to strike and the right of workers to be involved in setting up and monitoring safe systems at the workplace. Employers' liability to compensate for industrial accidents will also be considered.

Since the emphasis of this chapter will be on the duties which the employer owes by virtue of the contract of employment, it will be largely concerned with the employer's relationship to those who are working for him as his direct employees under a contract of employment between him and them. Where relevant protective employment legislation – which will normally be that which modifies the contractual relationship – also imposes upon an organisation duties which are not confined to those who are working for that organisation under a contract of employment, this will be clearly stated. This is likely to occur where a contractor is given an overriding responsibility to provide a system which encompasses a whole workforce, regardless of the contractor's exact relationship with individual members of that workforce. For example, S4 of Norwegian Regulations relating to worker protection and the working environment on fixed platforms of 1 June 1979 states:

> The principal enterprise shall ensure that all employers are informed of the provisions of the Working Environment Act and regulations issued pursuant thereto, and shall give information concerning the provisions to the necessary extent. The principal enterprise shall plan travel and work routine in such manner as to allow conformity with the provisions. The principal enterprise shall supervise safety and shall to the necessary extent exercise supervision to ensure that the provisions are complied with by the other employers.

The principal enterprise is *prima facie* the licensee; this is in keeping with the general trend of Norwegian shelf legislation which normally places an overall responsibility on the licensee to ensure the regulatory system is observed on the installation.

The boundaries between those matters which ought to be left to be freely negotiated and those which ought to be the subject of protective legislation are not always clear cut, and in any case tend to move over the course of time as social standards change. At any point of time 'welfare issues' tend to occupy a grey area between health and safety matters, which are generally perceived as being the appropriate subject for protective legislation, and those conditions of employment which may without hardship be left to be freely negotiated. There is, however, a tendency for what are initially regarded as welfare issues to be drawn towards the mainstream of protective legislation as advances in medical knowledge and improvements in standards of living widen the boundaries of protective legislation.

Thus environmental factors, which might in an earlier age have been deemed to be peripheral to welfare, are now – in recognition that a poor environment is likely to create stress – likely to be seen as being of significance to health. In this study, accommodation standards have been dealt with under the heading of health, because that seemed to be the most appropriate place to consider them. This is partly because so many of the regulatory provisions relating to accommodation deal with issues such as heating, lighting, ventilation and cleanliness, which clearly have a bearing on the health of a resident workforce; but partly also because sub-standard accommodation, which is not actually so poor as to cause disease, may be deemed in the light of modern medical science, to create stress factors harmful to health.

It is not entirely a one-way process, however, as the change in attitudes within the British sector to restricting hours of employment demonstrate. In the 19th century, when employment conditions were poor, it was understandable that control of the length of the hours of employment of women in factories was built into Factories Acts as a necessary aspect of protection of the health of the employees concerned. Now that working days are generally relatively short, these traditional protective measures have been deemed unnecessarily paternalistic and in contravention of equal opportunities legislation. In Britain, hours of work for men have never been the subject of general control.

Perhaps it is something more fundamental than a matter of translation from one language to another that this same distinction between welfare and health and safety issues is not apparently made in the Norwegian regulatory system.

Welfare issues

British legislation

That the distinction between health and welfare issues is never an easy one to make is often reflected in British protective legislation. The British Factories Act 1961 (which is not applied offshore) considered, under the Part devoted to welfare, such matters as the provision of drinking water, the provision of seating for workers during the course of employment, and accommodation for outdoor clothing.

The more recent Health and Safety at Work, etc Act 1974, in S2(1) imposes a general duty on an employer '... to ensure, so far as is reasonably practicable, the health, safety and welfare[2] at work of all his employees'. However, it is by no means clear what the addition of the word 'welfare' adds to the duty. It is interesting perhaps that the duty which the employer owes under S3 to those who are not in his employment relates only to exposing them to 'risks to their health or safety', and makes no reference to welfare. It may be questioned what, if any, difference the omission of the word 'welfare' makes to the nature and extent of the statutory duty which the employer offshore owes to his own direct employees and that which he owes to the employees of sub-contractors.

Certainly the limited welfare duties which are contained in the Offshore Installations (Operational Safety, Health and Welfare) Regulations 1976, made under the Mineral Workings (Offshore Installations) Act, are not imposed purely for the benefit of those who are the direct employees of the operator. The duties in question are contained in Regs 28 and 29. Regulation 28 prohibits the employment on offshore installations of persons under 18 years of age; Reg 29 requires an accurate record to be maintained of the hours worked by every person on the installation, although it does not place any restriction on the total number of hours which may be worked. Both these Regulations, which relate to the system of operation of the installation, are imposed primarily on the installation manager and require him to accept this degree of responsibility for whomsoever may be working on the installation regardless, of with whom that person's contract of employment has been made.

These examples serve to illustrate an important theme which runs throughout British labour law and is found in most other systems: the distinctions and overlaps between the obligations which the employer owes to or for the protection of those who are 'at work' in circumstances which are of general benefit to his organisation, and those which he owes to or for persons who are working for him under a contract of employment. This distinction may be inherent in a system of employment relationships based on freely negotiated contracts, but it is interesting that the concepts and philosophy of contract have been accepted and adopted in British statutory provisions which draw distinctions between what the law imposes on an employer for the protection of his own workers, and what it expects him to do in respect of workers who are not in his direct employment.

The distinction made between the duty which an employer owes to his employees under S2 of the Health and Safety at Work, etc Act, and that which he owes under S3 of that Act to those who are not in his employment, is a typical example of the way in which British law tends to impose somewhat stricter duties on an employer for the protection of those who are in his employment.

The contract of employment

In most legal systems the basis of the working relationship between an employer and a worker is a legally binding agreement. The major differences between legal systems relate to the extent to which the state leaves the contracting parties free to make their own terms and conditions for employment. The general experience of the past century or more has been of the increasing restriction of contractual freedom in the interests of protecting workers from exploitation by supposedly strong and ruthless employers.[3]

In this matter, unlike most of the issues which have been considered up to now, the offshore labour contract is likely to be little more than either an extension of the onshore regime of the coastal state or – in circumstances where there is little legislative protection given within the coastal state – a 'freely negotiated' agreement based on the philosophy of the domestic law of the state where the employer's principal office is situated.

In those cases where the petroleum activity is on a continental shelf where the coastal state has a well-developed system of protective employment legislation, the major questions will be as to the extent to which the onshore regime has been extended offshore – it is unlikely that there will have been an offshore regime developed independently by separate legislation. For this reason the authors will be drawing almost exclusively on British employment law, contrasting this with the laws of other coastal states only to the extent that their employment laws are known to be extended or reflected in particular petroleum provisions.

The development of onshore employment protection began in Britain with legislation clearly designed to protect health and safety, and was largely concerned with imposing duties on the enterprise for the protection of workers, without express reference to the contract of employment itself. However, although the legislation required minimum standards for the workplace, it was primarily concerned with workplaces where the contract of employment would be executed. By contrast legislation which, for example, restricts hours of work or imposes minimum rates of pay, clearly impinges on matters which are likely to be of the essence of the contract of employment itself.

As suggested, if carried to its logical conclusion, the distinction between protective legislation which controls the workplace, and that which controls the contract, is an imperfect one where it is the employee's workplace that is protected: if a person is expressly employed to work at a particular machine which lacks the guards required by safety legislation, the employer may be purporting to make the employee's acceptance of the breach of protective legislation a term of the contract of employment. Such attempts to avoid safety standards are, however, beyond the scope of this chapter.

Formation of the contract

British legislation

British law does not normally place restrictions on the persons with whom an employer may make a contract of employment; nor does it normally require a contract of employment to be made in any particular form. However, both these statements are subject to certain exceptions.

As has been noted above, a contract cannot lawfully require the employment offshore in the British sector of a person under the age of 18,[4] and the Aliens Restriction (Amendment) Act 1919 imposes nationality requirements in respect of the captain and other personnel employed to work on a British-registered ship.

These problems raise interesting issues of the possibility of conflict between the legality of the contract and the legality of its performance. To build on the two restrictions on performance outlined above by way of example: there is no illegality in an organisation which is engaged in petroleum activities on the British continental shelf or elsewhere, entering into a contract with a person under the age of 18 or an alien – the illegality would only be in requiring the young person to go offshore, or the alien to captain a British ship. The perpetration of the illegal act would not necessarily invalidate the contract, but, according to the regulatory provisions, one or both of the parties to the contract might incur criminal or other penalties.

By way of exception to the rule that contracts of employment need not in British law be formally made, the formation of merchant sailors' contracts have, since the principal Merchant Shipping Act of 1894, been subject to statutory control.[5] However, the requirements of this legislation with regard to this, and other aspects of the contract of employment, are somewhat marginal to the present review since they are part of flag rather than petroleum law, and will be applicable only to a British-registered ship which is incidentally employed in petroleum activities. They will, however, apply in whatever area of the world that ship is operating; but only to the ship's, as opposed to the drilling, crew.

Much more recently, general onshore employment protection legislation has required that an employer provide a full-time employee (as defined in the legislation) with written particulars of his terms of employment, but this statement is evidence of the contract rather than the contract itself[6] and does not have to be provided until the worker has reached the thirteenth week of full-time employment.

While this provision, like much of the Employment Protection (Consolidation) Act 1978 in which it is contained, is not intended for the protection of employees working outside Great Britain, the Act provided[7] that it might be extended offshore by Order in Council to the British sector of the continental shelf and to cross-boundary fields[8] and this power has now been exercised.[9] Many of the provisions of this Act will be referred to in subsequent parts of this chapter, since they are relevant to the rights of the employee during the performance, and in connection with the termination, of his contract of employment.

Norwegian legislation

As has been noted,[10] Norwegian law has specifically imposed restrictions on the employment of foreigners on her continental shelf, as part of her immigration legislation, and other nations such as Canada, have done or may well do likewise as they develop their submarine resources. As indicated above when outlining British law, the validity of any contract, which by its terms was in contravention of such a law, would depend on the rules of the court in which it was sought to enforce the contract. The contract might not be devoid of all legal effect but it would certainly not be capable of lawful execution within the coastal state which imposed the law which the contract violated.

The employment contract of offshore workers in the Norwegian sector of the continental shelf, except in so far as merchant shipping legislation governs the employment of sailors on Norwegian flag ships, is largely governed by the same Act of 4 February 1977, relating to worker protection and working environment, as governs the employment of labour onshore. The Royal Decree of 1 June 1979, which extends the onshore legislation to the offshore petroleum industry, indicates the extent of the extension offshore.

The framework legislation imposes no restrictions on the way in which the contract of employment is formed, but it and other legislation[11] do impose considerable restrictions on the terms of the contract; for example, hours of employment are tightly governed in Norway. However, violation of such statutory provisions is another instance in which the contract's validity would not be affected but sanctions might be incurred. Interestingly, however, while this Act provides a general code of employment protection, it makes no provision for recording the particulars of the contract in writing. It does, however, provide[12] that the employer shall not place discriminatory terms in the offer of employment.

Terms of contract and working conditions

However brief the contract of employment, it is likely to stipulate the general nature of the task for which the employee is engaged: the employer offering work of the nature specified[13] in the contract and the employee holding himself out as having the skill and competence to do the work offered. If the worker does not have the skill and qualifications he professes then, in most legal systems, he will not be entitled to retain the job, and this is likely to be so even in legal systems which grant general protection against dismissal.

It will, of course, be for the operator of the installation to ensure that sufficient numbers of persons are engaged and that those persons are of the right calibre to enable him to operate within the protective employment laws and to provide the safe systems of work which are required by the particular regime within which he is operating, even in instances where there are no regulatory standards for personal skills and qualifications for the tasks to be undertaken.[14]

Another element of the employment relationship is that the employee is expected to carry out the reasonable instructions of his employer. In the offshore environment, however, the personnel on the installation are subject to the authority of those on whom the law imposes responsibility for the safe management of the installation. The hierarchy of responsibility for the installation may well cut across the contractual relationship between the employee and his own employer.

Thus, in the British sector general responsibility for monitoring that the installation is functioning safely rests with the installation manager[15] and he has an overriding responsibility to prevent conduct which is unsafe or contrary to the specific requirements of Regulations made under the Mineral Workings (Offshore Installations) Act. In the last instance he has a statutory authority to keep under physical restraint any person whom he has reasonable cause to believe is endangering the installation.[16]

Responsibility in the Norwegian system ultimately rests with the licensee. In these circumstances it may frequently be that the employee who disregards the authority of the installation manager or the licensee's representative, as the case may be, will be in breach of his contract with his own employer, but there might be situations in which this was not the case.

In practice, statutory standards and safe systems are likely to be obtained and maintained through enforcement of contractual conditions by the head contractor stipulating the standards and methods of performance to be achieved by his contractors: employers engaged as contractors will then place appropriate terms in individual contracts of employment. It must be clearly understood, however, that although such contractual arrangements are a practical way of ensuring that the law is kept and safety achieved, they are not a way of discharging legal responsibility should breach of contract result in breach of the regulatory requirements.

British legislation
British employment protection legislation has not made any attempt to codify the contract of employment. Merchant shipping laws have given some protection to sailors concerning the payment of their wages, and the Wages Act 1986, which may be applied offshore by Order in Council, governs the manner in and extent to which an employer may make deductions from wages.[17] The Employment Protection (Consolidation) Act has provided a statutory floor of rights to employees during the currency of their employment contract. These rights are, however, largely concerned with trade union membership and activities, and entitlement to maternity leave. Up to the present these rights have been of little relevance to offshore employment because the workforce has not been heavily unionised and the tradition has been not to employ women offshore in the British sector of the North Sea; moreover the Sex Discrimination Act has yet to be applied offshore although there is now an intention to extend it to the British continental shelf.

Paragraph 5 of the Schedule to the Mineral Workings (Offshore Installations) Act enables Regulations to be made to place, 'Limits on hours of employment in any specified operation or in any specified circumstances', but this provision has not been implemented: the Operational Health, Safety and Welfare Regulations, as noted in Chapter 10, only require that a record be kept of hours worked. Nor has British merchant shipping legislation, up to this time, imposed any requirements on this matter.[18] It is unlikely, therefore, that there will be a need for there to be any express or implied contractual restriction on hours of work imposed, by virtue of British protective legislation, on those employed in the British sector.

Norwegian legislation

By contrast, the Norwegian regulatory system imposes strict controls upon both the number of hours worked in each day and the total number of hours worked in the course of a year.

Regulation of hours and duty periods for persons employed on fixed platforms are based on, but are modified version of, the general onshore requirements of the Worker Protection and Environment Act and are contained in the Regulations relating to worker protection and working environment of 1 June 1979.

Section 9 of these Regulations state that ordinary working hours shall not exceed 12 hours per day and 36 hours per week on an average over a period of a year. However, these provisions may be varied by trade union agreement, provided governmental consent is obtained. Section 9a introduces an important relaxation of the statutory restrictions, by providing that travelling time to and from the place of employment before and after each working period or stay period shall not be considered as working time. Clearly, this is very important to the operator in view of the length of the journey between shore base and installation; it is, however, controversial from the safety point of view, since it can mean that a worker is already tired before his working day begins.

Section 10 stipulates that rest breaks shall be included in the working day and these shall be of at least one hour per day in an eight-hour working day and one-and-a-half hours in a 12-hour working day. Section 11 provides that total working hours, including overtime, shall not exceed 16 hours per day. Total overtime must not exceed 200 hours per calendar year unless otherwise agreed in accordance with S50 of the onshore Working Environment Act: this enables the workers (through their workplace 'working environment committee'[19]). To decide that those who are willing to undertake extended overtime may work up to 300 hours overtime during the calendar year.

Section 11 of the Regulations states that normally the period of stay offshore must not exceed 14 days, but the employer and the elected worker representatives may agree to offshore employment for up to 16 days. In special circumstances, and following discussions with worker representatives, individual workers may have their period of stay extended for up to a further five days. The Petroleum Directorate may in special circumstances agree periods of stay somewhat greater than these.

Section 12 requires that there shall be an off-duty period of at least eight hours between two working periods. The onshore period between stays offshore must be at least one-third of the last period offshore.

Section 13 requires the principal enterprise on the installation to submit to the

Petroleum Directorate – preferably at least two weeks before operations commence – a plan showing the working time systems and periods of stay to be employed. This plan must show the working arrangements for all the contracting and sub-contracting firms as well as for the enterprise submitting the plan.

Hours of work on Norwegian mobile installations have hitherto been largely governed by the Regulations on working hours made by Royal Decree of 19 August 1977 pursuant to an Act of 3 June 1977 on working hours on board ships, and therefore these Regulations are of particular relevance to the ship's crew on board the installation, as opposed to the petroleum workers.

Regulation 3 of these Regulations requires that regular working hours shall be in accordance with a shift plan which has been established in advance and set for a certain period, based on the requirements of the installation and the petroleum employees. Regular working hours may be up to 12 hours a day but not exceeding an average of 42 hours a week over a maximum period of 12 weeks, and not exceeding an average of 36 hours a week over a 12-month period, or some alternative scheme collectively agreed.

Regulations 4 to 7 deal with payment systems, extra work needed for safety reasons, and overtime. Regulation 5 permits certain specified additional work to be carried out, mainly to deal with emergencies. Apart from this, overtime is not normally permitted and records must be kept of when it is done. Where these Regulations apply, Regulations of 11 May 1978 require a journal to be kept of hours worked.

Working hours for drilling crews were originally governed by Regulations of 29 August 1975. These Regulations have been repealed but the repealing instrument – the Royal Decree of 24 June 1977 – retained S6 of the earlier Regulations, concerning hours for drilling crews. However, it would now appear that the Royal Decree of 1 June 1979, which now governs the extension of the Worker Protection and Working Environment Act to petroleum activities, is wide enough in its intention to apply the Working Environment Act's provisions concerning hours to all those working on mobile installations on the Norwegian sector of the continental shelf.

The Regulations of 19 August 1977 would not appear to be applicable to foreign-registered mobiles, since they are 'flag' standards, and the ship's crew of a foreign mobile installation would, in any case, presumably be bound by any relevant employment protection provisions of the merchant shipping laws of the state where the installation is registered as a ship. It seems rather unlikely that the provisions of the Working Environment Act would be directly applied to the foreign drilling crew of such an installation, but the Norwegian authorities might well wish to be satisfied as to the working conditions of both ship's and drilling crew before granting approval for the installation to operate in Norwegian waters.

The provisions of the Regulations on the licensee's internal control would seem to be wide enough to enable the Petroleum Directorate to obtain and consider the work records of the persons employed on the installation as relevant matters within Reg 7, which deals with requirements associated with some of the activities of the licensee. It appears likely that the Directorate would expect labour conditions on the installation to be broadly comparable with those pertaining on a Norwegian mobile installation; and in fact, S2 of the Royal Decree of 1 June 1979 does provide for the exceptional application of the Act to foreign mobile installations.

Sickness and injury

Contracts of employment, particularly those which require employees to work in hazardous occupations, frequently contain express provisions concerning hospitalisation, compensation and income maintenance, following work-related accidents, or illness. It is also increasingly common for states to provide statutory schemes for income maintenance for their nationals during incapacity due to accident or illness. Generally, such state schemes are dependent on the employee having a sufficient contribution record at the time of his incapacity. It is also fairly common for protective employment legislation to contain safeguards against arbitrary dismissal of employees during sickness.

British legislation

The Social Security and Housing Benefits Act 1982 imposes on British employers the duty to pay statutory sickness pay, which is normally less than contractual pay, to qualified employees. This legislation would not, however, automatically apply to those who are engaged in petroleum activities, but S22 provides:

> The Secretary of State may make regulations modifying provisions ... in such manner as he thinks proper, in their application to any person who is, has been or is to be –
> (a) employed on board any ship, vessel, hovercraft or aircraft;
> (b) outside Great Britain at any prescribed time or in any prescribed circumstances; or
> (c) in prescribed employment in connection with continental shelf activities.

The Statutory Sick Pay (Mariners, Airmen and Persons Abroad) Regulations 1982 have brought persons employed in the petroleum industry on the British sector of the continental shelf within the definition of employees to whom the Act applies.

Only a careful consideration of the employee's contract in the context of the sickness or injury can determine whether it would be a breach of contract for the employer to dismiss the incapacitated employee. British law has, where the employer has not expressly reserved the right to dismiss in the case of ill health, traditionally considered the length and severity of the illness in relation to the length of the previous employment.[20]

Moreover, protective employment legislation does not necessarily protect even those who have onshore employment from dismissal during sickness. No employee has statutory protection unless he has two years' continuous employment with his employer before he falls sick. Whether it is fair for an employer to dismiss a long service employee who is sick depends upon whether it is reasonable for the employer to dismiss him bearing in mind the needs of the employer's business and the medical prognosis for the employee.[21] The position of British merchant sailors who fall ill during the course of a voyage is governed by the Merchant Shipping Acts.[22]

In common law jurisdictions it is frequently open to those who suffer personal injury to bring an action for damages. The English law of tort entitles an injured person to bring an action against any person whom he can establish owes him a duty of care. In order to win his action the accident victim must show that his injury (and this includes ill health) was caused by the negligent conduct of the person he is suing. The worker may sue, *inter alia*, either his own employer or another enterprise. While some of the states in the USA have personal injury law which may in some ways be more favourable to the victim than is that in Britain, employer's liability is restricted in the USA by the statutory workmen's compensation schemes which are usually the only avenue by which the injured workman can obtain compensation from his own employer. He may, however, have a common law right of action against some other person or organisation.

Whether a worker can sue at all, or obtain substantial compensation for injuries which have occurred outside the jurisdiction of the courts in which he wishes to bring his action, raises complex questions of international law. Some states are more disposed to entertain such actions than are others and some states give more substantial awards than do others. There are cases which have held that an employee of an English company can sue in an English court for compensation for personal injury suffered abroad in the course of his employment.[23]

Some of the problems of determining jurisdiction and application of law in cases involving private international law, were addressed in relation to the British continental shelf by the Continental Shelf Act 1964 S3. Provision has been made for the detailed amendment of this section by Ss22 and 23 of the Oil and Gas (Enterprise) Act 1982, but these amendments have yet to be activated. They will not, in any case, affect the general concept which is to enable civil actions to be brought in the relevant courts in Britain in respect of personal injuries suffered on offshore installations in the British sector of the North Sea.

Regulations have subsequently extended offshore the onshore requirement that an employer be insured to cover such liability to pay damages as he may incur as a result of employees suffering personal injury in the course of their employment as a result of the employer's negligence.[24] There might possibly be occasions when an accident victim whose injury occurred on the British sector of the shelf might prefer to sue in another jurisdiction. If the courts of that jurisdiction were prepared to hear his case he would be entitled to take this course.[25]

Norwegian legislation

In Norway, sickness benefits for nationals were, until a statute of 13 February 1976, entirely governed by the relevant social security legislation. Since the Act of 1976, an injured worker may claim compensatory damages under a system which is largely related to the concepts of negligence (including that the employer may be made vicariously liable for the wrongdoing of his employee). In cases where the worker has been exposed to risks which are over and beyond 'normal occupational risks', liability may be strict. There has been relatively little litigation; most cases are settled out of court. Trade unions may negotiate better terms than a court would order. For example, the widows of those killed in the Alexander L Kielland catastrophe received NOK800,000, whereas the normal court awards vary from NOK50,000–200,000.[26]

The Worker Protection and Working Environment Act gives considerable protection against dismissal during sickness. Section 64 provides:

> An employee, who is absent from work owing to accident or illness
> may not be given notice for that reason during the first six months after becoming unfit for work...

This provision is relevant to employment on fixed platforms on the Norwegian continental shelf and to drilling crews on other installations. Ships' crews on Norwegian mobile installations are protected by Regulations issued by Royal Decree of 19 December 1980 extending the provisions of the Seamen's Act to such vessels.

Termination of employment

A contract of employment may expire at a time, or upon an occasion agreed by the parties to it, either at the time when the contract was made or because by mutual consent they wish to bring it to an end, although there was no provision for this in the contract initially. More frequently, however, it will be brought to an end by the unilateral action of one of the parties, contrary to the wishes of the other.

British legislation

The last two decades have seen important developments in British labour law in the direction of giving employees security against arbitrary dismissal. The employee who has been continuously employed in full-time employment with the same employer for two years is normally entitled not to be unfairly dismissed[27] or made redundant without compensation,[28] within the statutory definitions of these concepts.

The requirement of continuity of service may well disentitle many offshore workers from these statutory protections. Moreover it is possible to avoid the legislative protection by giving a written contract in which the employee agrees that his engagement is only for a fixed term: the fixed-term contract may well be suitable for drilling or construction crews who are likely to be needed for a limited period only.[29]

However, the development of unfair dismissal law has done much to encourage employers to institute fair disciplinary proceedings. A dismissal for misconduct is 'fair' within the legislation provided that the employer has acted reasonably in dismissing in the circumstances.[30]

The severity of the misconduct (or a pattern of minor incidents with warnings) and whether the employer has conducted a proper factual inquiry are matters which will be weighed in the balance by the industrial tribunal, which is the statutory forum to which the worker may apply for a review as to whether the employer's dismissal of him for misconduct was fair.

It is significant that safety issues have frequently featured in dismissal cases. Unsafe conduct can certainly be considered misconduct warranting dismissal[31] but, on the other hand, an employer may be deemed to have 'constructively dismissed' an employee if he has required that employee to work in unsafe or unhealthy conditions.[32] If an employee satisfies an industrial tribunal that he has been unfairly dismissed his employer will either be ordered to take him back into employment[33] or will be required to pay compensation, the sum payable being related to the employee's wage, length of service and age.[34]

An employee who has not the requisite continuous service may nevertheless be able to claim against his employer under the legislation if he can show that he was dismissed for an 'inadmissible reason'. An inadmissible reason is one of the specified circumstances in relation to membership of or activities in connection with a trade union.[35] This particular protection could be a valuable tool for the trade union movement in its endeavours to increase its strength offshore in the British sector of the North Sea continental shelf.

British merchant shipping laws give their own special protection to sailors on ships registered under the British flag: traditionally this has given them tenure of employment only for the voyage or time contract for which they were engaged.

Norwegian legislation
The Norwegian Worker Protection and Working Environment Act gives employees stronger protection against dismissal than does the British law outlined above. It applies to fixed platforms. Section 60 states:

> Employees may not be given notice unless this is warranted by circumstances connected with the enterprise, the employer or the employee.

Dismissal on grounds of redundancy is, as in Britain, only warranted if the employer has no other suitable work to offer the employee. There appears to be no qualifying period of employment. It is envisaged by S61 that disputes between employer and employee concerning dismissal will be settled by negotiation and that legal proceedings will be a last resort. Section 62 provides for the situation where the dispute does go to court and states:

> When the court finds the notice [to terminate the contract] unwarranted and the employee so requires, the notice shall be ruled invalid. Nevertheless, in special cases, when so claimed by the employer, the court may decide that employment shall be terminated when, after weighing the interests of the parties, the court finds it clearly unreasonable that employment should continue.

Section 66 entitles the employer to terminate the contract summarily in the event of the 'employee's gross breach of duty or other material breach of the employment contract'.

While on the face of it the Norwegian law has much in common with the British, it is understood that because of the role of the working environment committees (see below) it is much less common for dismissals to occur, and if they do occur they are less likely to lead to litigation than is the case in Britain.

Worker involvement

British legislation

There is no general right for employees to be consulted by their employer concerning workplace arrangements onshore in Britain. There are only two exceptions to this and both apply only to workplaces where there is a recognised trade union. Moreover, whether a trade union is recognised depends on whether the union is sufficiently powerful to persuade the employer to recognise it; there is no statutory mechanism for making an employer recognise a trade union. A trade union therefore is only deemed to be recognised where the employer has entered into a recognition agreement with the union, or recognition can be implied from the conduct of the employer. The trade union must also be independent, ie not financially dependent on the employer for its existence. Where there is an independent recognised trade union, it is entitled to be consulted before the employer makes redundant any employee of the description in respect of which the trade union is recognised.

Onshore, by Regulations made under the Health and Safety at Work, etc Act[36] a recognised trade union is entitled to appoint safety representatives to consult with the employer on safety issues. These Regulations have not been applied offshore. Interestingly, merchant sailors do have some right to be consulted by virtue of the Merchant Shipping (Safety Officials and Reporting of Accidents and Dangerous Occurrences) Regulations 1982.

Norwegian legislation

Those working offshore in the Norwegian sector of the North Sea or on Norwegian-registered mobile installations do have considerable rights to be consulted by their employer on employment conditions. Regulations for safety delegates and working environment committees made by Royal Decree on 29 April 1977 give rights to onshore workers to elect safety delegates with the duty of furthering the implementation of the provisions of the Working Environment Act. Their duties expressly include evaluating and submitting opinions and studying reports on occupational health and safety matters. Enterprises which employ over 50 employees are required to have a working environment committee with an equal number of representatives from both the employer and the employees. This committee is both a decision making and an advisory body whose duties include preparing programmes on health and safety.

Regulations of 1 June 1979 for the extension of relevant parts of the Working Environment Act to fixed platforms and certain other offshore installations, recognise the right to have safety delegates and working environment committees offshore. Section 8 of these Regulations also enables procedures to be agreed under which a safety delegate may have the right to have dangerous work halted. Under the procedures, the safety delegate is required to ask management to halt the work. When a demand is made for work to be halted, the Petroleum Directorate must immediately be informed.

Regulations concerning protection supervisors and protection and environment committees on board ship laid down by the Directorate of Seamen on 15 November 1976, give similar rights to those which are enjoyed on fixed platforms to sailors employed on ships flying the Norwegian flag. These rights can be enjoyed by the ship's crew of a Norwegian ship which is operating as an offshore installation. The Regulations of 13 September 1985 do not apply to Norwegian sailors but may be applied to those workers (for example the drilling crew) on Norwegian-registered mobile installations who are not covered by the merchant shipping laws. Moreover they may be applied to cross-boundary fields outside the Norwegian sector of the continental shelf.

The Regulations of 1 June 1979 state that the Working Environment Act will not be applied to foreign mobile installations on the Norwegian sector of the continental shelf unless this is specifically stipulated by the Ministry through Regulations or individual decisions. It has already been suggested that operation in accordance with Norwegian standards might well be made a condition of approval of a foreign installation under the Royal Decree on licensee's internal control. It seems not unlikely that the same approach might be taken to the matter of worker involvement.

Norwegian Regulations of 19 March 1982 relating to safety manning in the event of an industrial dispute, require the licensee to ensure that an agreement is entered into with the workers' organisations relating to safety manning, in order to ensure that sufficient personnel remain on duty in the event of an industrial dispute. Workers are generally also required to carry out necessary safety work in accordance with agreed procedures before taking part in an industrial dispute. These Regulations appear to apply to all installations operating on the Norwegian sector of the continental shelf.

References

1   See *Priestley* v *Fowler* and the Factory Act, RWL Howells *26 Modern Law Review* 1963, p371.
2   Authors' italics.
3   Evidence may suggest that it is marginal employers, lacking the resources to provide good employment conditions, who are the real offenders. It is possible that the major oil companies operating offshore are, on this basis, better employers than are some of the smaller companies to whom they sub-contract servicing arrangements.
4   Norwegian law imposes the same minimum age requirement for fixed platforms. See Regulations relating to worker protection, etc of 1 June 1979: requirements for mobile installations are contained in Regulations of 21 October 1976 which apply the Seamen's Act to mobile installations.
5   Ss113–124 as amended by Merchant Shipping Acts 1950 S2 and 1970 S1.
6   Now S1 of the Employment Protection (Consolidation) Act 1978.
7   S137(2) and (5) and the Employment Protection (Offshore Employment) Orders 1976–81), but see also Oil and Gas (Enterprise) Act 1982 S37 Schedules 3 and 4 which prospectively substitutes for sub-sec 2 and prospectively repeals sub-sec 5 as from days to be appointed.
8   See also Employment (Continental Shelf) Act 1978.
9   The powers under the section have been exercised by the Employment Protection (Offshore Employment) Order 1976 SI No 766 as amended by the Offshore Employment (Amendment) Orders 1977 SI No 588, and 1981 SI No 208.
10 See Chapter 10.
11 Eg Regulations on working hours on board Norwegian drilling platforms and other mobile installations of 19 August 1977 below.
12 S55A. The section refers to political, religious or cultural views but makes no reference to racial or sexual discrimination. The British Sex Discrimination Act and Race Relations Act

have not yet been extended offshore; but S10(2) of the Sex Discrimination Act prohibits discrimination on a ship registered in Great Britain.

13   It is doubtful whether English employment law gives an employee any entitlement to be provided with work of the nature which he was employed to do; *Collier* v *Sunday Referee Publishing Co Ltd* [1940] 2 KB 647. The worker would, however, be entitled to refuse to do work which he was not employed to do, particularly if it were unsafe; *Ottoman Bank* v *Chakarian* [1930] AC 277.

14   See Chapter 10.

15   Mineral Workings (Offshore Installations) Act 1971 S5(2).

16   Ibid, S5(6) and (7).

17   See Merchant Shipping Act 1894 S131 and Merchant Shipping Act 1970 S7.

18   Regulations are currently in preparation.

19   See below.

20   *Storey* v *Fulham Steel Works* (1907) 24 TLR 89.

21   See *Spencer* v *Paragon Wallpapers Ltd* [1977] ICR 301.

22   See Merchant Shipping Act 1970 S25.

23   See *Matthews* v *Kuwait Bechtel Corpn* [1959] 2 Q.B.57 and *McDermid* v *Nash Dredging and Reclamation Co Ltd* [1986] 3 WLR 45; see *Spilada Maritime Corporation* v *Consulex Ltd, The Times* Law Report 24 November 1986 for guidelines on choice of forum.

24   The Employers' Liability (Compulsory Insurance) General Regulations 1971; the Offshore Installations (Application of the Employers' Liability (Compulsory Insurance) Act 1969) Regulations 1975.

25   The case of a Spaniard who, suffering injury on a ship in an English harbour, chose to sue his American employer in an American court received some press publicity.

26   Conference paper presented by Professor Asbjorn Kjanstad of Oslo University to XI International Conference of the International Society of Labour Law and Social Security in Venezuela, September 1985 (Conference Papers Theme III, Volume 2).

27   Employment Protection (Consolidation) Act 1978 S54.

28   Ibid S81.

29   Ibid, S55(2)(b) and S142.

30   Ibid S57.

31   Eg *Alidair Ltd* v *Taylor* [1978] ICR 445.

32   *British Aircraft Corp Ltd* v *Austin* [1978] IRLR 332.
Under S55(2)(c) a constructive dismissal occurs when 'the employee terminates that contract ... in circumstances such that he is entitled to terminate it without notice by reason of the employer's conduct.'

33   Ibid, S69. The order may be for either 'reinstatement' or 're-engagement', ie restoration to his former job with all seniority, etc or given a job on the terms dictated by the court.

34   Ibid, S72. In addition there may be a compensatory award to take into account loss of seniority, etc.

35   Ibid, S58. Dismissal relating to trade union membership.

36   The Safety Representatives and Safety Committees Regulations 1977.

13

Damage Control

Principal legislative provisions

British legislation

Radioactive Substances Act 1960.
Employers' Liability (Compulsory Insurance) Act 1969.
Merchant Shipping Act 1970.
Mineral Workings (Offshore Installations) Act 1971.
Health and Safety at Work, etc Act 1974.
Petroleum Act 1987
The Employers' Liability (Compulsory Insurance) General Regulations 1971.
The Offshore Installations (Registration) Regulations 1972.
The Offshore Installations (Logbooks and Registration of Death) Regulations 1972.
The Offshore Installations (Inspectors and Casualties) Regulations 1973.
The Offshore Installations (Public Inquiries) Regulations 1974.
The Offshore Installations (Application of the Employers' Liability (Compulsory Insurance) Act 1969) Regulations 1975.
The Offshore Installations (Operational Safety, Health and Welfare) Regulations 1976.
The Offshore Installations (Emergency Procedures) Regulations 1976.
The Offshore Installations (Life-saving Appliances) Regulations 1977.
The Health and Safety at Work, etc Act 1974 (Application outside Great Britain) Order 1977.
The Submarine Pipe-lines (Inspectors etc) Regulations 1977.
The Offshore Installations (Fire-fighting Equipment) Regulations 1978.
Control of Lead at Work Regulations 1980.
Ionising Radiation Regulations 1985.
The Classification, Packaging and Labelling of Dangerous Substances Regulations 1984.

Norwegian legislation

Act of 22 March 1985 pertaining to petroleum activities.
Act of 4 February 1977 relating to worker protection.
Regulations concerning safety relating to the Act of 22 March 1985 made by Royal Decree of June 28 1985.
Regulations concerning the licensee's internal control made by Royal Decree of 28 June 1985.

Regulations for the Government Action Control Group in the event of accidents leading to or involving risk of extensive pollution laid down by Royal Decree of 19 November 1982.

Regulations concerning labelling etc of chemical substances and products which may involve a hazard to health.

Regulations concerning safety manning in the event of an industrial dispute stipulated 19 March 1982.

Regulations concerning stand-by vessels stipulated 28 December 1983.

Regulations on arrangements on and below deck and for safety measures, etc stipulated 31 January 1978 and 13 January 1986.

Regulations on life-saving appliances on fixed installations stipulated 8 February 1978.

Regulations for life-saving appliances on drilling platforms stipulated 3 February 1982 and 13 January 1986.

Regulations on approval of survival suits stipulated 10 November 1980 and 13 January 1986.

Provisional Regulations for living quarters on production installations of 2 April 1979.

Regulations for mobile drilling platforms with installations and equipment laid down 10 September 1973 and 13 January 1986.

Regulations on welding equipment on drilling platforms laid down 1 March 1977.

Regulations concerning fire extinguishing and fire safety measures on Norwegian drilling units of 31 January 1984.

Regulations concerning the mustering of employees on board ships laid down 16 June 1975.

Temporary Regulations for electrical installations on board drilling platforms stipulated 26 July 1985 and laid down 1 December 1974.

Regulations for the use of radioactive sources on board drilling platforms laid down 15 May 1973 and 13 January 1986.

The whole offshore regulatory regime has the objective of damage control; ideally this will be achieved through building safety into systems of operation so that accidents do not occur. The same approach to all industrial activity is becoming increasingly common: it is the 'Robens philosophy',[1] which in Great Britain received statutory recognition in the Health and Safety at Work etc, Act 1974. This approach is also to be seen in the Norwegian Worker Protection and Working Environment Act 1977 and in the recent Act of 22 March 1985 concerning petroleum activities.

The offshore industry magnifies the safety problems encountered onshore in several respects. First, there are particular dangers inherent in the activities of exploring for and exploiting of a volatile substance like petroleum; second, the operation has to be undertaken in a hostile environment; and third, the workforce has to reside at the workplace.

This book has so far been primarily concerned with safe operation, but since safe operation involves an ever-present awareness of the need to avoid accidents while being prepared for their consequences, much that has been said about safe systems of operation has also touched upon systems for response to catastrophe.

This chapter will bring together four matters particularly related to damage control, namely: 1) systems for hazardous activities; 2) accident reporting arrangements; 3) contingency planning; 4) Removal of installations.

Ultra-hazardous activities

Occupational health and safety legislation has developed historically by the identification of, and legislation for, particular hazards. In the course of time, this piecemeal approach to the regulation of working conditions tends to result in a 'checklist' rather than a systematic approach to safety. The 'systems approach' which is increasingly adopted by the developed world is reflected in the recent legislation of both Norway and Britain, but neither system has entirely abandoned the earlier approach of legislating particular standards or systems for especially significant hazards. The present system in both regimes is therefore one of placing on organisations and individuals general duties to devise and enforce safe systems of work, while requiring particular regulatory standards to be placed within these general systems.

This philosophy of combining the general requirements concerning safe systems with particular requirements for particular standards for particular situations has, in both regimes, been employed both onshore and offshore. Both offshore and onshore the particular regulatory standards are at the moment – in what may be described as an 'interim phase' in the development of regulatory legislation – to be found partly in regulatory codes which predate the new systems legislation but have not been repealed by it, and partly in new sets of Regulations which have been enacted since the new principal legislation, and consciously build on the 'systems approach' contained in the general Act.

The radical change in approach, which was established in Norway by the principal Act of 22 March 1985, has left the Norwegian regime with principal legislation which is general in form and intended to be the foundation for a very broad approach to safe systems, while for the time being retaining much detailed regulatory control from the former system. There is a somewhat similar situation in the British sector, where the detailed Regulations made under the Mineral Workings (Offshore Installations) Act are enforced, in many instances, side by side with the broad general requirements of the Health and Safety at Work, etc Act.

In both sectors, the exclusively offshore legislation has to be seen in conjunction with Regulations which have been made for the control of particular hazards onshore. As was noted in an earlier chapter,[2] the Norwegian state initially adopted a broader approach than Britain to the extension offshore of regulatory standards which were originally intended for domestic use. However, since the Health and Safety at Work, etc Act the British approach has been to regulate in general terms for the control of hazards at whatsoever workplace they may be found. Now that the Health and Safety at Work, etc Act has been extended to offshore activities, new British Regulations are increasingly likely to be framed for application to offshore as well as onshore activities.

In this section, some of the particular hazards which have been subject to special regulatory control offshore will be noted. It must, however, be borne in mind that this list cannot be exhaustive, in that in both systems, now that the emphasis is on the operator's general duties,[3] the inspectorate may well expect well-established onshore regulatory standards to be used in offshore operation, whether or not these standards are directly legally enforceable offshore.

British legislation

The Offshore Installations (Operational Safety, Health and Welfare) Regulations make express provision for the identification of hazardous areas and for safe systems of work within those areas. Regulation 2 provides:

> (1) There shall be included in the written particulars comprised in the operations manual relating to any offshore installation ... drawings of the installation clearly and accurately showing any part of the installation in which there is likely to be danger of fire or explosion from the ignition of gas, vapour or volatile liquid (in these Regulations referred to as a 'hazardous area'.
>
> (2) The door or hatch for any opening giving access to a hazardous area shall bear on the outside the words 'HAZARDOUS AREA' in red capital letters at least 50 millimetres high.

As noted already, the Regulation itself imposes no specific requirements for the protection of individuals or the control of work operations in relation to areas which are identified as hazardous. However, Reg 2 must be seen in association with the immediately following Regulations, namely Reg 3 concerning work permits and Reg 4 concerning dangerous substances.

Regulation 3 applies to the specific tasks of welding and flame cutting and any other work involving or giving rise to a source of ignition and work on electrical equipment. None of these tasks may be carried out by any person on any offshore installation without the written instructions of the installation manager. These instructions, given to a responsible person, must state the nature of work, the period during which the work may take place and any precautions to be taken to avoid endangering the safety of the installation and any persons on it. Schedule 2 sets out the matters which must be covered by written instructions, under the following general categories: drilling, production procedures, electrical procedures, mechanical equipment procedures, personal procedures, procedures to secure safety of the installation.[4]

Clearly, when these inherently dangerous tasks have to be performed in hazardous areas, the requirement for the institution and maintenance of a safe system of operation is more than usually great.

Regulation 4 provides for a safe system for the storage and use of substances whose specific characteristics make them dangerous:

> (1) No radioactive, corrosive, toxic or explosive substance or any substance which is stored or used at a pressure greater than atmospheric pressure, shall be kept by any person on an offshore installation except in suitable receptacles clearly marked with the contents at a place as far as reasonably practicable from any hazardous area and any living accommodation.
>
> (2) No flammable substance shall be so kept except in such receptacles at a place as far as reasonably practicable from any other hazardous area and any living accommodation.
>
> (3) Every place at which for the time being any substance mentioned in paragraph (1) and (2) above is kept shall be in the charge of a responsible person.
>
> (4) Without prejudice to Regulation 2(2) above, any door or hatch giving access to any

place at which for the time being any such substance as is mentioned in paragraph (1) or (2) above is kept, shall bear on the outside the word 'DANGER' in red capital letters at least 50 millimetres high with an adequate description or indication of the substance in question.

(5) No such substance as is mentioned in paragraph (1) or (2) above shall be used by any person on any offshore installation unless all reasonably practicable precautions have been taken against any danger to which any person on the installation may be exposed by the use of the substance.

(6) There shall be accurately shown on the drawings of the installation in the operations manual referred to in Regulation 2(1) above any part of the installation in which are stored any of the substances mentioned in paragraph (1) or (2) above.

In addition to these statutory systems for handling materials with dangerous properties, the Regulations made under the Health at Safety at Work, etc Act concerning the Packaging and Labelling of Dangerous Substances, the Control of Lead at Work and Ionising Radiation, all have special significance for activities offshore. Packaging and Labelling Regulations, which since they are made under an EEC Directive apply widely to goods manufactured in and supplied from Europe, give practical guidance in assisting the user to identify the dangerous properties of a large number of substances. Neither the Directive nor the Regulations actually apply offshore but if the supplier, as seems likely, observes the standards he would normally observe onshore, the employer who uses such substances at his workplace offshore will be put on notice that he should ensure that the substance is used as the supplier has advised, if he is to comply with the general duty which Ss2 and 3 of the Health and Safety at Work, etc Act impose upon him. The Lead and Ionising Radiation Regulations[5] have been expressly applied to offshore activities and impose their own particular requirements as to safe systems.

Regulation 2(1)(g) of the Offshore Installations (Inspectors and Casualties) Regulations 1973 empowers a Department of Energy inspector to:

> ... require the owner or manager or any person on board or near to an offshore installation to do or to refrain from doing any act as appears to the inspector to be necessary or expedient for the purpose of averting a casualty, or minimising the consequences of a casualty.

Similar powers are given to pipe-line inspectors under Reg 3 of the Submarine Pipe-lines (Inspectors, etc) Regulations 1977.

The powers given in the Health and Safety at Work, etc Act to serve prohibition notices must also be borne in mind, especially the provision for the service of a deferred prohibition notice to enable an operation to be run down rather than stopped immediately, in cases where this approach may create a lesser risk than stopping immediately.

Both these enforcement mechanisms are particularly likely to be invoked where there is a breach of specific regulatory requirements for the control of ultra-hazardous activities. They are therefore important weapons in the exercise of damage control.

There is also the ultimate sanction of revocation of licence. This may be regarded as a valuable deterrent held in reserve for any case of repeated, flagrant or reckless neglect of safety precautions.

As has been noted in the previous chapter, the Offshore Installations (Application of the Employers' Liability (Compulsory Insurance) Act 1969) Regulations 1975 extend to offshore installations the Employers' Liability (Compulsory Insurance) Act 1969 and the Employers' Liability (Compulsory Insurance) Act General Regulations 1971, which implement the Act itself. This legislation requires employers to have insurance cover to meet their legal responsibilities to pay damages to compensate those of their employees who suffer accidents in the course of their employment, as a result of the employer's negligence or breach of statutory duty. This legislation, of course, is only very indirectly related to the safe systems which might prevent the occurrence of accidents, but it assists in damage control in that it is intended to ensure that there is a fund from which compensation may be paid.

Norwegian legislation

Section 12.4 of the Act of 4 February 1977 relating to worker protection and working environment makes provision for the circumstances in which work involves safety hazards. After stipulating that premium wages must not be paid for work where this may materially affect safety, it continues:

b) If work is to be carried out in the enterprise that may involve particular hazard to life and health, a special directive shall be issued prescribing how the work is to be done and the safety precautions to be observed, including any particular instruction and supervision.
c) When the work is of such a nature that it involves danger of a disaster or disastrous accident, plans shall be drawn up for first aid, escape routes and rescue measures, registration of employees present at the workplace and so on.
d) Employees shall be informed of the regulations and safety rules, etc relating to the area concerned and of the plans and measures mentioned under c).

These provisions apply to work offshore as well as on land in the Norwegian sector, in addition to the special regulation of hazardous situations which is provided for in the petroleum legislation. Section 25 of the Regulations concerning safety made by Royal Decree on 28 June 1985 requires the licensee to prepare an organisational plan at each phase of offshore activity, including exploration drilling. The plan must state the number of personnel and their location in the various areas of work.

The modern system under the Regulations for licensee's internal control provides scope for the Petroleum Directorate to require plans both for the designation of hazardous areas and for the organisation of work systems to control the risk from especially dangerous activities. A systematic approach to hazards is to be found in the Guidance for design of fixed platforms which is linked to the Regulations for structural design. Together they form the framework for the licensee's systematic approach to risk management. This conceptual framework is to be seen in conjunction with detailed requirements, such as the provisional Regulations for living quarters on production installations of 2 April 1979 which require, in S2.2.1, that living quarters must be so situated that they are securely separated from dangerous areas and activities.

Regulations on arrangements on and below deck laid down on 31 January 1978 (shelf) and 13 January 1986 (flag) are more specific in the systems they require for hazardous work. Section 12 dealing with precautions during special work, like the British system, requires written instructions to be issued in relation to the use of welding equipment, although there does not appear to be the same dependence on a permit-to-work system. This section also provides systems for the use of high-pressure spraying gear and for work in tank rooms, tunnels, etc where there may be an insufficiency of oxygen, or poisonous or explosive gases.

Section 12 also requires that a plan shall be established for every drilling unit, showing open areas, closed rooms and tanks, and indicating where there are special risks of explosion or an unhealthy (eg insufficiency of oxygen) atmosphere. The section also deals with the cleaning and repair of containers which have been used for inflammable, explosive or poisonous matter.

There are also special Regulations of 21 March 1977 concerning the use of welding equipment on Norwegian drilling platforms, and Regulations of 13 January 1986 concerning radioactive substances used on Norwegian drilling platforms.

The Norwegian system contains similar provisions for the labelling of dangerous substances as are to be found in British law, but the Norwegian Regulations in the Royal Decree of 26 November 1982 are more directly applied offshore than is the case with the British equivalent.

Apart from the general provision that violations by individuals of specific Regulations may result in criminal prosecution under the Penal Code, repeated or serious disregard of the Norwegian Regulations may result in forfeiture by the licensee of his licence. Prosecutions for breach of shelf legislation will in future be brought under the Petroleum Act rather than the Penal Code.

Accident reporting

Most occupational health and safety regimes include a system for reporting of accidents and dangerous occurrences. Apart from enabling records to be kept on the safety performance of the industries concerned, and of individual organisations within the industry, such records enable the identification of particularly hazardous activities. Post-accident investigation enables not only the punishment of those, if any, whose wrongful actions or omissions have caused the accidents, but also enables positive work to be done to set up safe systems to eliminate the hazards and prevent the recurrence of accidents in the future. Indeed, it is noteworthy that in the British sector the system of accident reporting, investigation and inquiry not only developed piecemeal, but also preceded the system of emergency procedures which is intended to reduce and control the loss which accidents can entail.

Those incidents which have relatively minor consequences are normally dealt with by inquiry and report within the enforcement agency,[6] which may result in prosecution by the agency itself, or through another agency such as the police force, in systems where the inspectorate has no inherent power to prosecute. Major catastrophes are not infrequently followed by public inquiry and an official published report. The offshore petroleum industry is subject to this form of regulatory control. There have been a number of official reports and inquiries, which are referred to by reference to the name of the installation involved in the catastrophe, for example: 'The Sea Gem',[7] 'The Ekofisk',[8] 'The Alexander Kielland',[9] and 'The Ocean Ranger'.[10]

British legislation

The general system of accident reporting which now applies onshore in Britain for the reporting of accidents and dangerous occurrences, set out in the Reporting of Injuries, Diseases and Dangerous Occurrences Regulations 1985 made under the Health and Safety at Work, etc Act, does not apply to offshore installations. The situation offshore for operations is governed by the Offshore Installations (Logbooks and Registration of Death) Regulations 1972, the Offshore Installations (Inspectors and Casualties) Regulations 1973 and the Offshore Installations (Public Inquiries) Regulations 1974.

Reporting of accidents which occur on British-registered ships is governed by the Merchant Shipping Act 1970. The Submarine Pipe-lines (Inspectors), etc Regulations 1977 impose duties in respect of the reporting of accidents in relation to pipe-lines and pipe-line activities.

The Logbooks and Registration of Death Regulations apply to installations whether or not they are registered as vessels and required to keep another logbook in that capacity. Regulation 2 requires that in respect of other installations an installation logbook must be maintained on every offshore installation at all times when the installation is in controlled waters,[11] except that in the case of a fixed installation under construction or in the course of assembly or dismantlement, it will be sufficient to maintain the installation logbook on an attendant vessel. Among the information which must be recorded in the logbook, Reg 3(2) requires:

(b) adverse weather conditions, collisions, structural changes and major repairs, surveys and other occurrences relevant to the safety, seaworthiness or stability of the installation;

(c) safety drills, accidents and injuries to persons, and the occurrence of disease and death;

(d) emergencies and apprehended emergencies and measures taken to meet or avoid them, whether relating to the installation or to personnel;

Every entry in the logbook must be made (and signed by the installation manager) within 24 hours of the time at which the occurrence the subject of the entry took place or as soon as practicable thereafter, and must include the time at which the occurrence took place and the time at which the entry was made and signed.

Regulation 8 requires the registration of deaths and persons lost. It provides:

Where any person –
(a) dies on an offshore installation or is lost from an installation in circumstances such that it is reasonable to believe that he has died; or
(b) dies in or on a lifeboat, liferaft or other emergency survival craft belonging to an offshore installation or is lost therefrom in such circumstances as aforesaid; or
(c) otherwise dies or is lost in such circumstances as aforesaid in the neighbourhood of an offshore installation while engaged in any operation connected with the installation;

unless the death is one which will have to be reported under British merchant shipping law; for this purpose a 'return of death' must be made by completing Part I of the special statutory form. The completed form must be sent by the installation manager to the owner of the installation as soon as is practicable, and in any event within 10 days of his becoming aware of the death or loss to which the return relates. On receiving the form the owner must, within 10 days, complete Part II of the form and send it to the Registrar General of Shipping and Seamen. Regulation 10 requires the employer of the dead or missing person to notify the next of kin.

Part II of the Inspectors and Casualties Regulations provides a system for the reporting of any casualty to the Secretary of State for Energy to enable its investigation. 'Casualty' includes incidents involving danger to, as well as loss of, life and the Regulations apply to casualties involving persons working from attendant vessels in the course of any operation undertaken on, or in connection with, an offshore installation, as well as casualties involving those actually employed on, or working from, an installation.

It might be persuasively argued that dangerous occurrences, where there is no risk of injury, do not have to be reported, especially as under similar Regulations applying to pipe-lines, dangerous occurrences are fully described. It may well be that, now that there is a new code for onshore reporting, the opportunity will be taken in the foreseeable future to revise the offshore provisions.

Regulation 9 requires that where a casualty has occurred the manager of the offshore installation on or near to which it occurred shall:

... in the most expeditious manner practicable, immediately inform the owner of the installation of its occurrence with brief particulars of the casualty, including the position of the installation, the time of the casualty and the identity of any person killed, lost or seriously injured;

He must also sign an entry in the logbook and, within three days after the occurrence of the casualty, provide the installation owner with specified written particulars of the accident. As soon as he is informed, the owner must give all the information he has to the Secretary of State: he must put this information, together with other specified particulars, in writing within three days.

Currently the form in use for this is OIR8 for a fatality and OIR9 for a serious injury or occurrence. He must also, as soon as he learns that any person injured in a casualty has died, give notice of his death to the Secretary of State, even though he may also be required to send a return of death to the Registrar General of Shipping and Seamen in accordance with other Regulations such as those set out above. If an investigation is not planned the inspector gives permission for the site to be disturbed so that normal working can be resumed.

Regulation 11 requires that the place where the casualty occurs shall not be disturbed until the inspectorate has had an opportunity to carry out an investigation. That is to say it must not be disturbed until after three days from the time when the owner has informed the Secretary of State of the casualty or the inspector has concluded his investigation, whichever occurs first.

The Offshore Installations (Public Inquiries) Regulations 1974 enable the Secretary of State to order that a public inquiry may be held into any casualty or other accident involving loss of or danger to life which has occurred on, or in connection with, the presence of or operation of an offshore installation in waters in the British sector. The Regulations do not apply to registered vessels which at the time of the casualty are in transit to or from a station.

Accidents, injuries and diseases which disable persons from working for more than three days have to be submitted in like manner as a consolidated list once every three months the current form being OIR10. A disease for these purposes is any disease of body or mind and not necessarily, as is the case onshore, an occupational disease. (Incidentally, in the Norwegian system a similarly broad definition of disease is used.)

The Secretary of State is empowered to appoint a competent person to hold a 'court' of inquiry. Regulations 6 and 7 give the court wide powers to board and inspect installations and vessels and take evidence from persons. These powers do not appear to be limited to the circumstances immediately surrounding the accident which led to the setting up of the court.

The court is required to report to the Secretary of State:

> ... stating fully the circumstances of the casualty or other accident and the opinion of the court as to the cause together with any observations and recommendations which the court thinks fit to make with a view to the preservation of life and the avoidance of similar accidents in the future.

Regulation 5 of the Submarine Pipe-lines (Inspectors, etc) Regulations 1977 requires the owner of a pipe-line in controlled waters to report certain occurrences in relation to any pipe-line or pipe-line works to the Secretary of State for Energy. Specified information relating to the occurrence, including the nature of the occurrence, the time and place of the occurrence, and the identity of any person dead or seriously injured must be provided in the most expeditious manner practicable. Within five days written particulars must be provided in accordance with Schedule 2 of the Regulations. The occurrences to which these requirements apply are set out in Schedule 1 and include numerous situations concerned with property damage as well as: 'Any accident causing death or loss of, or serious bodily injury to, any person.'

Where an accident causing serious bodily injury has been reported and the person injured dies within 12 months after the occurrence of the accident, the owner of the pipe-line or the proposed pipe-line must, as soon as it comes to his knowledge, give notice of the death to the Secretary of State for Energy.

In addition, Reg 6 requires the owner of the pipe-line to make a formal return to the Secretary of State for Energy, once every three months, of every injury or disease suffered by any person (including any diver) while engaged in pipe-line works. He must also report any discovery of any defect in any plant or equipment intended to form part of the life-support system of a diver, or non-compliance with instructions given to persons engaged in diving operations for such works which might endanger the safety or health of a diver.

These special regulatory provisions must be put in the context of the Fatal Accidents and Sudden Deaths Inquiry (Scotland) Act 1976 which makes general provision for the police to hold inquiries into fatal accidents which occur in that part of the continental shelf to which Scottish jurisdiction applies. Similar duties to investigate and report on sudden deaths are given by common law to coroners in England.

Norwegian legislation

Section 21 of the Act of 4 February 1977 applies to fixed platforms in the Norwegian sector of the North Sea. This provision does not, however, apply to the ship's crew on Norwegian drilling vessels who are subject to Norwegian maritime law, nor does it normally apply to foreign mobile structures.

Section 21 requires that an employer shall report injuries and diseases. It provides:

> If, as the result of an occupational accident, an employee loses his life or is seriously injured, the employer shall immediately, and by the quickest possible means, notify the Labour Inspection and the nearest police authority. The employer shall confirm his notification in writing. The safety delegate shall receive a copy of the confirmation. The Directorate of Labour Inspection may also require such notification to be given in other cases.
>
> The Directorate of Labour Inspection may require that the employer shall report to the Labour Inspection:
> a) occupational accidents in respect of which notification is not required under the first paragraph or second paragraph, also including acute poisonings, and any near accidents (serious accidents or acute poisonings that are averted).
> b) any disease that is, or may be caused by the work or by conditions at the workplace.
>
> Regulations concerning the extent and the implementation of the duty to give notification of, and the duty to report, injuries and accidents shall be issued by the Directorate of Labour Inspection.

The Directorate of Labour Inspection has delegated its responsibility on the continental shelf in these matters to the Petroleum Directorate. Section 22 imposes a similar duty to report upon any medical practitioner who attends an employee suffering from an occupational injury as defined by the National Insurance Act, or any other disease which the medical practitioner believes is due to the employee's work. This duty also applies to company doctors who learn of diseases in the course of their work. It is perhaps arguable that this section is wide enough to impose an obligation on a medical practitioner even if the accident or illness in question was suffered by a person working on an installation or pipe-line not covered by the Act itself.

As far as fixed platforms are concerned the only other reference to accident reporting would appear to be in S7 of the Regulations concerning the licensee's internal control made by Royal Decree of 28 June 1985. This Regulation includes a requirement that the licensee shall ensure:

> ... that there (within the working environment) is a continuous control and mapping of the working environment in the activity with regard to the risks, health risks and welfare aspects, and that the necessary corrective measures are implemented.

This would appear to be tantamount to a reminder that the licensee must ensure that Ss21 and 22 are complied with. It is also interesting to note that the Regulations concerning internal control are not limited in their application to fixed platforms. They also require in S5 that, as a prerequisite to obtaining consent to operate in Norwegian waters, the licensee must provide assurance that he will comply with the Norwegian Petroleum Directorate's lawful requirements concerning documentation.

Sections 5 and 7 together may constitute sufficient assurance that the licensee can be held responsible for ensuring the reporting of accidents, even those which occur on foreign mobile installations. The delegation of enforcement powers by the Labour Inspectorate to the Petroleum Directorate means that in practice the relevant accident reports will be made to the Petroleum Directorate rather than to the Labour Inspectorate itself.

Specific provision is made for accident reporting in relation to petroleum activities on the Norwegian sector of the North Sea continental shelf by S17 of the safety Regulations of June 1985. However, flag ships will clearly have to comply with their own flag law, which is likely to require the reporting of accidents. Flag law will certainly be a means of ensuring that accidents on Norwegian-registered mobiles are reported within the Norwegian jurisdiction.

Additionally, it must be borne in mind that accident investigation is a matter for the police as well as the Petroleum Directorate in the Norwegian sector. While the Petroleum Directorate always investigates to find out what action should be taken, this investigation is often carried out in parallel with that carried out by the police. However, the purpose of the Petroleum Directorate is not prosecution. The role of the police is apparently greater than is the case in the British sector: the Norwegian shelf legislation has applied the Penal Code and police powers in relation to it right across the Norwegian sector of the continental shelf. In the Norwegian system major catastrophe inquiries are normally conducted by the police also and – certainly when the Norwegian petroleum regime was more clearly one of approval of the licensee's activities by the Petroleum Directorate – an accident inquiry put the Petroleum Directorate itself under scrutiny to a greater extent than would be the case in a similar post-accident situation in the British sector.

Contingency planning

Ensuring that an installation is prepared to meet any emergency – particularly one which may endanger life – is a shared responsibility of government agencies and the operator. Legislation may stipulate standards for equipment, such as survival suits, and systems for ensuring persons are ready and trained for emergencies. Responsibility must rest with the industry for ensuring that the statutory standards are complied with and that a safe system is set up and maintained to enable a rapid and suitable response to any emergency.

In both the British and Norwegian regimes the regulatory systems for contending with emergencies have been developed in the light of experience – some of it gained, unfortunately, from the accident reports submitted by the industry itself and also from the experience of the merchant shipping industry, which has for centuries contended with some of the same hazards, namely those associated with evacuation from the installation to the hostile maritime environment.

In the case of the operation of the offshore petroleum industry, the likelihood of evacuation proving necessary is arguably greater than in the case of a ship engaged merely in navigation because, to the natural hazards created by the action of the elements upon the structure of the installation, there must be added the hazards of fire and explosion created by the physical properties of the petroleum which is being exploited.

The response of the legislatures in both regimes has been to stipulate standards for the provision of systems of evacuation of personnel from the installation, and to require the industry to devise and record with the enforcement agencies its systems for contending with emergencies in drilling and production – such as wellhead control, blow-out and shut-down procedures. The operator is also required to satisfy the inspectorate that he has a satisfactory system for communicating the existence of an emergency to a shore base which is, in its turn, equipped to assist on the scale which may be necessary. Catastrophe arrangements will need to provide for the involvement of public services in rescuing personnel and protecting property; measures to control pollution may be second only in importance to the saving of human life.

British legislation

The expression 'contingency planning' is the accepted translation of the Norwegian approach to catastrophe arrangements. The British legislation in the narrow context uses the expression 'emergency procedures'. The Norwegian expression is adopted here, however, because the emphasis, much of which has already been noted, throughout the whole planning and execution of the British system, is to reduce the possibility of emergency arising by building into the normal systems of operations a surplus capacity to meet extraordinary stresses, and by isolating and controlling hazardous features of day-to-day operation. This approach of building into procedures a capacity for withstanding stress is to be found particularly in the Offshore Installations (Construction and Survey) Regulations 1974 and the Offshore Installations (Operational Safety, Health and Welfare) Regulations 1976.

When applying for a Certificate of Fitness under the Construction and Survey Regulations, the owner must, in accordance with Reg 5(1)(e), submit an operations manual, and the installation will be evaluated bearing very much in mind the environment in which it will be operating. The whole emphasis of Schedule 2 of the Regulations is that the criterion of fitness for the installation is whether is is 'capable of withstanding any foreseeable combination of forces'. Special consideration must, however, be given to escape routes, stability control (ballast pumping) and the provision of an emergency power supply.

The Operational Safety, Health and Welfare Regulations set down systems for the control of certain daily routine operations which, if not carefully carried through, could result in catastrophe. Some of these have already been considered in this chapter under the heading of hazardous activities; others, such as helicopter landing and radio communications, were considered in Chapter 8. Making special provision for these activities not only reduces the likelihood of their resulting in catastrophe but, in some cases, sets in operation the preliminary stages of response to catastrophe. For example, the work-permit system should focus attention on the operation for which the permit is granted and alert the installation manager to monitoring the progress of the activity to its safe completion. The involvement of the radio operator in the helicopter landing exercise not only enables routine instructions and advice to be exchanged, but enables emergency systems to be put into operation with the minimum delay should this prove necessary.

These two regulatory codes are therefore very important aspects of contingency planning for damage control in its widest sense. The statutory requirements for the provision of mechanisms to ensure an adequate response to catastrophe are contained in three sets of Regulations: the Offshore Installations (Emergency Procedures) Regulations 1976, the Offshore Installations (Life-saving Appliances) Regulations 1977 and the Offshore Installations (Fire-fighting Equipment) Regulations 1978. In terms of setting up safe systems, the first of these is by far the most important, the other two being primarily concerned with standard of provision rather than system of operation.

A most important aspect of the Emergency Procedures Regulations is the duty which they impose upon the owner of the installation and the concession owner to provide an emergency procedure manual. The manual must make provision for certain specified emergencies, and it must provide information about the general systems which will be employed for damage control in the event of an emergency arising. The manual must contain the names and addresses of any public or local authorities to whom any particular emergency is to be reported and specify the method of, and the time for, making the report. A copy of the emergency procedure manual (in English and any other language which is appropriate) must be made available at a suitable and readily accessible place on the installation. The Secretary of State must be supplied with a copy of the manual, if he so requests, and, if he is of opinion that the manual does not make sufficient provision for securing safety in the event of an emergency, he may serve on the owner of the installation a notice stating his opinion and specifying the matter for which different arrangements ought to be made, and requiring the manual to be amended accordingly.

The emergencies to which the manual must give particular consideration are:

(a) a fire or explosion;
(b) a blow-out of a well;
(c) a leak or spillage of any oil or gas;
(d) a storm or severe weather conditions affecting the stability of the installation;
(e) a movement of the sea-bed affecting the stability of the installation;
(f) a failure of the structure of the installation;
(g) a failure of the equipment of the installation affecting the safety of persons;
(h) a failure of the means for keeping the installation on station;
(i) a collision involving the installation;
(j) an accident involving a helicopter;
(k) a person falling from the installation; and
(l) a death, a serious injury or illness.

The emergency procedure manual must set out:

(a) a code of signals suitable for transmission by means of a general alarm system for signifying the occurrence of specific emergencies and the action to be taken in respect of any of them, and instructions for the transmission of any of those signals;
(b) instructions for the operation of the emergency equipment;
(c) instructions for rendering safe any work being carried on or any equipment being used;
(d) instructions for evacuating all persons on or near the installation therefrom; and
(e) the place on the installation at which plans showing the location of the emergency equipment are to be provided.

The manual should also contain particulars of:

(a) the emergency services arranged for divers;
(b) the action to be taken by the stand-by vessel;[12]
(c) any action arranged by the owner of the installation to be taken by any person on land or another installation for providing assistance; and
(d) any available search and rescue services.

In addition, the Regulations make provision for muster lists for all installations which are manned by five or more persons. The muster list must state, in respect of every person on the installation, the station to which each person shall go in the event of an emergency, and any duties to be carried out in an emergency which are assigned to particular persons.

The muster list must set out the person appointed to be in charge of each emergency station and the persons to whom duties are assigned in connection with the following matters:

(a) the closing of wells;
(b) the closing of pipes carrying hydrocarbons or other flammable substances and the associated valves and vents;
(c) the closing of fire doors and ventilators;
(d) the closing of watertight doors;
(e) the stopping of machinery;
(f) the extinction of fire;
(g) the equipping of survival craft and liferafts and making them ready for use;

(h) the launching of survival craft and liferafts; and

(i) the securing of the safety of visitors to the installation.

Regulation 10 of the Emergency Procedures Regulations, which was cited in Chapter 8, requires that there must be present within five nautical miles of every installation a stand-by vessel ready to give assistance in the event of an emergency on or near the installation. The stand-by must be capable of accommodating on board all persons who may be on the installation at any time, and must be equipped to provide first-aid treatment for such persons. The agreed standards, which are not legislative, for equipping of stand-bys, are to be found in the *Green Book*.

Under the same Regulations, all persons on an installation must be made aware of their duties when musters and drills take place, and persons with responsibilities in the event of any emergency must be trained in the duties assigned to them.

The Life-saving Appliances Regulations stipulate the type of life-saving equipment which must be available. They require that survival craft, liferafts, lifebuoys and lifejackets must be of a type approved by the Secretary of State. They also require that all life-saving appliances must be available for immediate use and, where necessary, protected from damage. There must be plans on the installation showing the position of all life-saving appliances. Appliances must be examined at least once every two years by a person authorised by the Secretary of State.

The Regulations require that every normally manned installation must be provided with approved, totally enclosed motor-propelled survival craft having, in the aggregate, sufficient capacity to accommodate safely on board all persons on the installation. They also require that there be further provision of one of the specified types of survival craft to ensure that the installation has, in total, enough capacity to accommodate safely on board one-and-a-half times the number of persons on the installation.

It is further provided that there must be suitable devices for lowering the survival craft into the water and disengaging the craft from the launching apparatus. The survival craft must also be equipped and provisioned to the standards set out in the Regulations. The alternative is to provide survival craft with a capacity for all on board, plus liferafts to ensure accommodation for twice the number of persons on the installation.

The installation must be provided with lifebuoys in such numbers and stowed in such places that at least one is readily accessible from any part of a deck of the installation from which a person is liable to fall into the water. Every installation must also be provided with at least as many lifejackets as one-and-a-half times the number of persons on the installation.

The installation must have suitable and sufficient means for persons to descend from the installation to the water in an emergency. Finally, the installation must have a general alarm system capable of raising the alarm by signals audible at every part of the installation where aural communication is practicable, and similarly it must have a public address system capable of being heard distinctly at all parts of the installation where persons are frequently present and aural communication is practicable. Both the general alarm and the public address systems must be capable of providing a conspicuous visible warning in every part of the installation where persons are present and aural communication is not possible.

The Fire-fighting Equipment Regulations stipulate the type of fire-fighting equipment with which installations must be provided. They require, in general, that any fire-fighting equipment which is provided on an offshore installation must be of a type for the time being approved by the Secretary of State for installations of a class or description which includes that installation. They require that every installation must be provided with equipment suited to the place at and circumstances for which it is to be installed: thus there are particular requirements for accommodation areas.

The equipment which may have to be provided, according to the circumstances, includes an automatic fire detection system; an automatic gas detection system; a manually activated fire-alarm system; fire mains; hydrants and fire hoses; water deluge and water monitors; an automatic sprinkler system; a fixed fire extinguisher system; fire extinguishers; four sets of fireman's equipment; fire blankets; special fire-fighting equipment for the helicopter landing area and remote control safety equipment for certain specified purposes.

Norwegian legislation

There is in the Norwegian system the same emphasis as in the British on the control of hazardous activities in order to minimise the risk of catastrophe. In that system, as in the British, it is required that installations shall be classified in terms of explosion risk and divided into zones according to the degree of risk; electrical installations must be properly designed, used and maintained and installations must be so laid out as to minimise the consequences of fire and explosion. In particular, Regulations have been made for mobile installations.[12] The licensee is required to ensure that all personnel on the installation exercise caution regarding anything which might cause fire.[13]

There are, however, specific requirements for contingency plans. Section 46 of the principal Act of 22 March 1985 lays down a general duty on the licensee at all times to maintain efficient contingency preparedness plans, with a view to countering accidents and emergencies which may lead to loss of lives or personal injuries, pollution or major damage to property. The section continues:

> The licensee is responsible for all necessary measures being initiated to prevent or minimise harmful effects, including necessary measures to return the environment to the condition it had before the accident occurred.

The section enables the Ministry to decide whether other parties shall make available necessary resources for contingency preparedness plans for the account of a licensee. The section is particularly relevant to the provision of major resources to counteract pollution.

Section 22 of the Regulations of 28 June 1985 on safety imposes a similarly general requirement:

> The licensee shall prepare safety contingency plans for installations and vessels which

participate in the activity. Particular attention must be paid to the safety of the individual. Each individual must be thoroughly familiarised with these rules.

Later, in S52 of the same Regulations, it is repeated:

The licensee shall prepare a contingency plan in case of accidents and dangerous situations as mentioned in S46 of the Act. The plan shall be continuously updated and shall be made known to the parties involved. The plan shall be submitted to the Ministry.

Section 53 stipulates that the plan must include:

(a) an organisational plan with precise description of responsibilities and reporting lines, and the responsibility of individuals in the event of accidents and dangerous situations as specified in S46 of the Act.

(b) a plan of equipment for fighting an actual accident or hazardous situation with a precise description of the nature and type of equipment, capacity, location, transportation method, correct usage and area of use.

(c) an action plan with a precise description of alarm and communication systems, including systems for notifying the authorities, the duties of individuals, when and how emergency equipment is to be used and how the operations shall be accomplished, measures for limiting the extent of the damage from the accident or hazard, and rules for winding up the operation.

Ensuring that these general requirements are complied with is done through the Regulations concerning the licensee's internal control, which require the licensee to satisfy the Petroleum Directorate on numerous matters before being granted consent to operate in the Norwegian sector. Among the requirements contained in S7 is:

... that an emergency preparedness system is established and maintained so that the necessary measures can be activated effectively and authorities involved notified, in case accidents or incidents involving danger should occur.

Separate Regulations made by Royal Decree of 19 November 1982 provide for a Governmental Action Control Group to be summoned into action in the event of major pollution. This group includes representatives of the Ministry of Environment, the Ministry of Labour, the Ministry of Trade and Industry, the Police and the Defence Command. In practice, the Ministry of Labour is likely to be represented by the Petroleum Directorate and the Department of Trade and Industry by the Maritime Directorate.

In addition to this regulatory framework there are detailed Regulations concerning stand-by vessels,[14] life-saving appliances,[15] and approval of survival suits.[16]

Provisions for emergency procedures on all mobile installations, whether or not they are Norwegian-registered flag ships, are contained in Regulations for mobile drilling platforms laid down on 10 September 1973 (shelf) and 13 January 1986 (flag). They include the usual maritime procedures for muster lists, musters and drills.[17]

Removal of installations

The salvage of wrecks after shipping disasters is a difficult operation and one which has received some publicity lately, both in respect of the lifting of the Tudor warship *Mary Rose* and the investigation of the wreck of the luxury liner, the *Titanic*. In less spectacular situations wrecks may lie on the bed of the sea for all time; indeed, there is some pressure for them to be left undisturbed if they are in fact graveyards. These same problems are associated with the loss of mobile installations and, again, considerable publicity was given to the difficulties encountered in righting the Alexander Kielland after it capsized and towing it back to the coast at Stavanger.

The Geneva Convention in Article 5.5 requires that: 'Any installations which are abandoned or disused must be entirely removed.' It is worth reflecting, however, that when the Convention was drawn up installations of the magnitude of those which now exist were not envisaged.

Nevertheless, there is a requirement for removal, albeit in a somewhat modified form in Article 60 of the Jamaica Convention. This provides that:

> ... installations which are taken out of service or which are no longer in operation shall be removed to protect the safety of shipping, due account being taken of generally accepted standards which have been stipulated in this instance by the relevant international organisation. In connection with removal, account should also be taken of fisheries, protection of the marine environment and the rights and obligations of other states. The depth, the position and the dimensions of installations which are not removed in their entirety shall be announced in appropriate manner.

This Article is not, of course, yet in full force.

The problems of removing a fixed installation, either as an aftermath of a major catastrophe or because it has exhausted its useful production life, are ones which have yet to be faced. British licences normally impose upon the licensee a formal requirement not to abandon a well without the consent in writing of the Minister, and a basic requirement for removal; but no proper consideration has been given to regulating for the removal of installations, beyond that the Offshore Installations (Registration) Regulations 1972 Reg 6 impose upon the owner of the installation the obligation to notify the Secretary of State if an installation is dismantled, abandoned or destroyed.

The Petroleum Act 1987 places a new, fairly general duty upon the owner to remove an installation after it has ceased to be operable for petroleum activities. The main thrust of this new law is that it enables specific Regulations to be made to stipulate more detailed requirements in respect of removals, and empowers the Secretary of State to require abandonment programmes from owners and other persons.

The principal Norwegian Act of 22 March 1985 hints in S63 that the licensee may have some responsibility for removal, in that it states that revocation of a licence, surrender of rights or lapsing of rights for other reasons does not lead to release from the economic obligations which follow from the Act, Regulations under it, or separate conditions. If a work commitment or other commitment has not been fulfilled, the Ministry may demand payment. It is interesting that the new Norwegian legislation, made in realisation of the scale of fixed platforms, no longer imposes upon the licensee the obligation to remove which was contained in Ss.17 and 21 of the Royal Decree of 9 July 1976.

It must be a matter of increasing concern, however, as the end of the useful life of the larger production installations inevitably grows nearer, that no real provision has been made for their removal, for the reason that it is by no means clear how the task can be undertaken. Simply to abandon them must be hazardous to others using the high seas, particularly as, in the course of time, as their condition deteriorated they would be likely to be less visible and might in due course be reduced to obstructions below rather than on the surface of the sea. However, to allow them to be transferred to other uses, such as bird sanctuaries, is no real solution since it can only prolong their existence for a limited time.

Not only might abandoned installations cause serious accidents but it must be questionable for how long the former licensees or owners could be held liable to compensate the victims. Insurance on the scale required could hardly be maintained indefinitely: not only would the cost be very high but the greater the passage of time, the greater the difficulty in tracing the insurer and the insured.

There is some evidence of an increasing awareness of this problem, in the press as well as in governmental circles, but no realistic solution has yet been proposed.[18]

References

1   Lord Robens chaired the seminal committee of the 1970s. See *Safety and Health at Work. Report of the Committee 1970–1972* HMSO, Cmnd 5034, 1972.
2   See Chapter 2.
3   In the Norwegian sector see Regulations concerning the licensee's internal control made by Royal Decree on 28 June 1985; in the British sector see the general duties (particularly Ss2 and 3) of the Health and Safety at Work, etc Act.
4   Insofar as these are maintenance matters they are discussed in detail in Chapter 10.
5   See Chapter 2. The Radioactive Substances Act was extended offshore by the Continental Shelf Act 1964.
6   The Department of Energy publishes an annual report, 'Development of the Oil and Gas Resources of the UK' known as the *Brown Book*, which provides, *inter alia*, statistics relating to accidents in the British sector. There is a similar publication by the Petroleum Directorate in Norway.
7   *Report of the Inquiry into the Causes of the Accident to the Drilling Rig Sea Gem* Ministry of Power, HMSO, Cmnd 3409.
8   *The Uncontrolled Blow-Out in the Ekofisk Field 22 April 1977* Report No 65 to the Storting (1977/ 78).
9   Ministry of Local Government and Labour, Norway: Report No 67 to the Storting (1981–82): *The Alexander L Kielland incident* Oslo, Norway.
10   *Royal Commission on the Ocean Ranger Marine Disaster – Canada Newfoundland and Labrador – Report One: The Loss of the Semi-submersible Drill Rig Ocean Ranger and its Crew – 1984.*
11   Ie (i) designated areas of the shelf; (ii) territorial waters; (iii) tidal waters – to the point where a river ceases to be tidal.
12   See Regulations on electrical installations on board drilling platforms of 11 November 1975 and Regulations òn arrangements on and below deck and for safety measures, etc of 31 January 1978 (shelf) and 13 January 1986 (flag).

13   Ss37–39 of the Regulations concerning safety made by Royal Decree on 28 June 1985.

14   Stipulated 28 January 1983. The Petroleum Directorate considers that the Alexander L Kielland disaster demonstrated the need for new stand-by regulations, and new guidelines on emergency preparedness are expected in 1987 (Norwegian Petroleum Directorate's *Annual Report* for 1985 at p81).

15   Stipulated for fixed installations 8 February 1978 and for drilling platforms and other mobiles 3 February 1982 (shelf) and 13 January 1986 (flag).

16   Stipulated 10 November 1980 (shelf) and 13 January 1986 (flag).

17   There are also separate Regulations concerning fire extinguishing and fire safety measures on drilling units of 31 January 1984 and Regulations concerning the mustering of employees on board ship of 16 June 1975.

18   See Death of an oil platform, *New Scientist* 27 February 1986; Scuppering the Rigs, *The Times* 14 July 1986; and One law for the rigs, *New Scientist* 15 January 1987.

Appendix 1

Legislation relating to the safety, health and welfare of persons working offshore on the North Sea continental shelf and elsewhere:

British legislation

Shelf legislation

Legislation applicable directly or indirectly to petroleum activities in the British sector of the North Sea continental shelf.

Submarine Telegraph Act 1885: protection of submarine cables, telephonic cables, high voltage cables, and latterly, pipe-lines.

Anchors and Chain Cables Act 1889: testing of cables: standards re-enacted in the Offshore Installations (Construction and Survey) Regulations 1974.

Petroleum (Production) Act 1934: licensing of petroleum activity onshore: applied offshore by the Continental Shelf Act 1964, qv.

Coast Protection Act 1949: provides, *inter alia*, for the lighting, identification and removal of obstructions to navigation; applied to installations by the Continental Shelf Act 1964.

Wireless Telegraphy Act 1949: regulates wireless telegraphy stations in UK; extended offshore by the Continental Shelf Act 1964.

Civil Aviation Act 1949: power to regulate air navigation by Order in Council; as amended by the Mineral Workings (Offshore Installations) Act 1971, applies to all aircraft on or in the neighbourhood of an offshore installation.

Radioactive Substances Act 1960: makes provision for registration and disposal of radioactive material; extended offshore by the Continental Shelf Act 1964.

Continental Shelf Act 1964: makes provision for the exploration for and the exploitation of petroleum resources within the sea-bed of the British continental shelf.

Anchors and Chain Cables Act 1967: makes new provision for the testing of anchors and chain cables; may be applied offshore by Regulations made under the Mineral Workings (Offshore Installations) Act.

Employers' Liability (Compulsory Insurance) Act 1969: makes insurance against certain civil claims for occupational injuries compulsory; applied offshore by Regulations made under the Mineral Workings (Offshore Installations) Act 1971.

Mineral Workings (Offshore Installations) Act 1971: makes provision for safety, health and welfare on offshore installations.

Misuse of Drugs Act 1971: restricts the possession, use and supply of controlled drugs; may be applicable offshore under the Continental Shelf Act.

Health and Safety at Work, etc Act 1974: health, safety and welfare of all persons at work; applied offshore by Order in Council in 1977.

Employment Protection Act 1975: makes provision for the notification of and consultation over redundancies; applied offshore.

Petroleum and Submarine Pipe-lines Act 1975: regulates the construction and use, including safety and inspection, of submarine pipe-lines in 'controlled waters'.

Fatal Accidents and Sudden Deaths Inquiry (Scotland) Act 1976: provides for holding inquiries into fatal accidents in connection with the exploration and exploitation of the sea-bed in that part of the British sector of the continental shelf to which Scottish law applies.

Employment (Continental Shelf) Act 1978: enables extension of certain domestic labour legislation to 'cross-boundary' oil fields.

Employment Protection (Consolidation) Act 1978: consolidates statutory rights of employees arising out of employment.

Oil and Gas (Enterprise) Act 1982: *inter alia*, amends the legislation relating to offshore safety, health and welfare.

Police and Criminal Evidence Act 1982: *inter alia*, clarifies powers of constables on offshore installations.

Housing and Social Security Act 1982: makes provision, *inter alia*, for the payment of statutory sick pay by employers.

Food and Environmental Protection Act 1985: lays down, *inter alia*, standards of food hygiene; applies to seas within British fishing limits.

Wages Act 1986: regulates the payment of wages; applies to offshore operations.

Petroleum Act 1987: makes provision, *inter alia*, for the establishment of safety zones and the removal of installations.

Regulations

The Continental Shelf (Designation of Areas) Orders 1964 SI no 697, *et seq*: designate areas within the UK continental shelf for purposes of exploration for and exploitation of petroleum resources.

The Petroleum (Production) Regulations 1966 SI no 898: contained, *inter alia*, model clauses for licences granted before 20 August 1976.

The Continental Shelf (Jurisdiction) Order 1968 SI no 892 *et seq*: divides the 'designated areas' into English, Scottish, and Northern Ireland areas for purposes of application of the appropriate law thereto.

Anchors and Chain Cables Rules 1970 SI no 1453: prescribe manner in which required tests must be carried out.

Employers' Liability (Compulsory Insurance) General Regulations 1971 SI no 117: detailed provision for compulsory insurance against employers' liability for occupational accidents and ill health.

The Offshore Installations (Registration) Regulations 1972 SI no 702: establish register of installations and impose duty to register upon owners of installations.

The Offshore Installations (Managers) Regulations 1972 SI no 703: appointment and termination of appointment of offshore installation managers.

The Offshore Installations (Logbooks and Registration of Death) Regulations 1972 SI no 1542: maintenance of installation logbook, personnel records, and returns of deaths.

The Offshore Installations (Inspectors and Casualties) Regulations 1973 SI no 1842: powers of inspectors, and duties of owners and managers of installations.

The Offshore Installations (Construction and Survey) Regulations 1974 SI no 289: require certification of installations, lay down standards of construction and equipment, and for the termination of certificates.

The Offshore Installations (Public Inquiries) Regulations 1974 SI no 338: provision for holding of inquiries into incidents involving loss of or danger to life, in connection with offshore installations.

The Offshore Installations (Application of the Employers' Liability (Compulsory Insurance) Act 1969) Regulations 1975 SI no 1289: extend the 1969 Act (qv) to employment in connection with offshore installations.

The Employers' Liability (Compulsory Insurance) (Offshore Installations) Regulations 1975 SI no 1443: provide for above.

The Continental Shelf (Protection of Installations) Order 1976 SI no 260 *et seq*: provides for safety zones in connection with offshore installations.

The Employment Protection (Offshore Employment) Order 1976 SI no 766: extends certain employment protection legislation to employment in designated areas.

The Submarine Pipelines (Diving Operations) Regulations 1976 SI no 923: made provision for diving in connection with a submarine pipe-line; superseded, for all practical purposes, by the Diving Operations at Work Regulations 1981 qv.

The Offshore Installations (Operational Safety, Health and Welfare) Regulations 1976 SI no 1019: detailed requirements for safety, health and welfare in connection with offshore installations.

The Offshore Installations (Emergency Procedures) Regulations 1976 SI no 1542: emergency procedures, manuals, muster lists, drills and stand-by services in connection with offshore installations.

The Petroleum (Production) Regulations 1976 SI no 1129 (now repealed): made provision for model safety clauses in licences issued before the coming into force of the 1982 Petroleum (Production) Regulations (qv). These clauses imposed, *inter alia*, a duty to comply with certain public health, food and drugs, and misuse of drugs legislation, otherwise not enforceable offshore.

The Air Navigation Order 1976 SI no 1783: regulates registered and other aircraft on or in the neighbourhood of an offshore installation.

The Offshore Installations (Life-saving Appliances) Regulations 1977 SI no 486 (*et seq*): provision and testing of life-saving appliances used in connection with offshore installations.

The Employment Protection (Offshore Employment) (Amendment) Order 1977 SI no 588: further extension of employment protection legislation to designated areas.

The Submarine Pipe-lines (Inspectors) Regulations 1977 SI no 835: grant powers to inspectors and impose duties upon owners and proposed owners of pipe-lines.

The Health and Safety at Work etc Act 1974 (Application outside Great Britain) Order 1977

SI no 1232: applies the 1974 Act to offshore installations, associated activities, and pipe-line works.

The Offshore Installations (Fire-fighting Equipment) Regulations 1978 SI no 611 (*et seq*): provision and examination of fire-fighting equipment, including operational instructions.

The Control of Lead at Work Regulations 1980 SI no 1248: makes provision for safe systems for protecting employees from the effects of lead at work; applicable to persons employed in petroleum activities offshore.

The Offshore Installations (Well Control) Regulations 1980 SI no 1759: make further provision for drilling operations.

The Diving Operations at Work Regulations 1981 SI no 399: provide for safe activities in diving at work; applicable to persons employed in offshore petroleum activities.

The Submarine Pipe-Lines (Electricity Generating Stations) Regulations 1981 SI no 750: make provision for regulation of electricity generating stations in connection with submarine pipe-lines utilised for North Sea petroleum activities.

The Petroleum (Production) Regulations 1982 SI no 1000: provide, *inter alia*, for a sole model safety clause in petroleum licences issued after 1982, to the effect that the licensee shall comply with any safety instructions given by the Minister in writing.

The Statutory Sick Pay (Mariners, Airmen, and Persons Abroad) Regulations 1982 SI no 1349: extend 'statutory sick pay' rights to employment in a designated area or a prescribed area.

The Submarine Pipe-lines Safety Regulations 1982 SI no 1513: safety in the construction and operation of submarine pipe-lines.

The Offshore Installations (Included Appliances and Works) Order 1982 SI no 1542.

The Social Security and Statutory Sick Pay (Oil and Gas (Enterprise) Act 1982) (Consequential) Regulations 1982 SI no 1738: amend SI 1982 no 1349 (above).

The Asbestos (Licensing) Regulations 1983 SI no 1649: further control of activities involving asbestos; applicable offshore.

The Offshore Installations (Application of SIs) Regulations 1984 SI no 419.

The Freight Containers (Safety Convention) Regulations 1984 SI no 1890 as amended: standards of construction and maintenance for certain freight containers; applicable offshore.

The Classification, Packaging and Labelling of Dangerous Substances Regulations 1984 SI no 1244: provide for safe labelling and packaging of certain dangerous substances; not applicable to petroleum activities offshore, but may affect such substances if delivered offshore from Britain.

The Asbestos (Prohibition) Regulations 1985 SI no 910: prohibition of activities involving certain types of asbestos; applicable to offshore activities.

The Ionising Radiations Regulations 1985 SI no 1333: protection of employees against ionising radiations at work; applicable to persons employed in offshore petroleum activities.

The (proposed) Offshore Installations and Pipe-line Works (First Aid) Regulations.

Codes of practice, notes of guidance, etc

Department of Energy, Petroleum Engineering Division: *Safety, Health and Welfare*

Instructions issued under the Petroleum (Production) Regulations (1964 onwards).

Department of Energy: *Continental Shelf Operations Notices.* (1972 onwards).

Department of Transport : *Standard Marking Schedule for Offshore Installations* (1976).

Department of Energy: The Offshore Installations Guidance on Design and Construction (1977).

Department of Energy: *The Offshore Installations Guidance on Life-saving Appliances* (1978).

Department of Energy: *Diving Safety Memoranda* (issued from 1978 onwards).

Health and Safety Executive: (Training of offshore sick-bay attendants) (Guidance Note MS 16; 1978).

Department of Transport: *Assessment of the suitability of stand-by vessels: Instructions for the guidance of surveyors* (1980).

Department of Energy: *The Offshore Installations Guidance on Fire-fighting Equipment* (1980).

Health and Safety Executive: *Offshore Construction (Guidance Notes)* (1980).

Department of Energy: *Training Guidance Notes:*

No 1 *Qualifications for Radiotelephone Operators on Offshore Installations* (1985).

No 2 *Guidance on the selection, training, and appointment of persons as Helicopter Landing Officers* (1985).

No 3 *Guidance in Training in Survival for Persons Working on Offshore Installations* (undated).

No 4 *Guidance on the selection and training of Ballast Control Officers* (undated).

No 5 *Guidance on the selection and training of Offshore Crane Operators* (undated).

Health and Safety Executive: *The Control of Lead at Work: Code of Practice* (1985).

Health and Safety Executive: *The Ionising Radiations: Code of Practice* (1985).

Health and Safety Executive: *The Work with Asbestos and Asbestos Coating: Code of Practice* (revised) (1985).

Flag legislation

Legislation applicable to UK-registered ships, also installations and other ship-like structures registered as ships, anywhere in the world. Thus regulates ships participating in petroleum activities on the North Sea or other continental shelves.

Merchant Shipping Act 1894: makes provision for a register of British ships, and for legislative powers governing such ships.

Public Health (Scotland) Acts 1897 and 1945: Provision for standards of hygiene in Scotland.

Merchant Shipping (Safety and Load Line Conventions) Act 1932: relates, *inter alia*, to load lines and life-saving equipment.

Public Health Act 1936: provision for standards of hygiene on, *inter alia*, British ships.

Merchant Shipping Act 1948: standards of crew accommodation, also certification of key members of crews.

Merchant Shipping (Safety Conventions) Act 1949: makes provision for safety of life at sea.

Merchant Shipping Act 1964: makes provision for construction, survey and certification of British-registered ships.

Merchant Shipping (Load lines) Act 1967: further provision relating to load lines.

Merchant Shipping Act 1970: power to make Regulations for safety, health and welfare for persons on British ships; for crew accommodation; and gives the Department of Trade power to inspect such ships 'outside the United Kingdom'.
Merchant Shipping Act 1974: regulates activities on or in submersible or supporting apparatus operating in or outside Britain.
Sex Discrimination Act 1975: prohibits discrimination on the grounds of sex in Britain; applied to employment on British ships.
Merchant Shipping Act 1979: further provision for making of Regulations concerning the safety, health and welfare of persons on British ships; also empowers the Department of Trade to register as United Kingdom ships '... things designed or adapted for use at sea ...'
Merchant Shipping Act 1984: empowers the Department of Trade to use Improvement and Prohibition Notices (qv) to enforce safety Regulations on British ships
Safety at Sea Act 1986: extends the powers in the Merchant Shipping Act 1979 to make Regulations concerning safety, health and welfare at sea.

Regulations
The Merchant Shipping (Life-saving Appliances) Rules 1965 SI no 1105: provision for life-saving appliances on British-registered ships, including mobile installations registered as ships; also support, supply, stand-by and accommodation vessels.
The Merchant Shipping (Load Line) Rules 1968 SI no 1453: load lines on British ships.
The Merchant Shipping (Radio) (Fishing Vessels) Rules 1974 SI no 1919: provision for radio-telephone apparatus on British fishing vessels; required on all stand-by vessels on British continental shelf.
The Merchant Shipping (Diving Operations) Regulations 1975 SI no 116: regulates diving operations from or near a British ship: but not diving operations to which the Diving Operations at Work Regulations (qv) apply.
The Merchant Shipping (Crew Accommodation) Regulations 1978 SI no 795: lay down standards of crew accommodation on British ships.
The Merchant Shipping (Code of Safe Working Practices) Regulations 1980 SI no 686: make provision for Code of Safe Working Practices for Merchant Seamen (qv).
The Merchant Shipping (Certification of Deck Officers) Regulations 1980 SI no 2026: make provision for qualifications of deck officers on British ships.
The Merchant Shipping (Certification of Marine Engineers) Regulations 1980 SI no 2025: makes provision for qualifications of marine engineers on British ships.
The Merchant Shipping (Safety Officials and Reporting of Accidents and Dangerous Occurrences) Regulations 1982 SI no 1876: make provision for appointment of safety officials and reporting of accidents and incidents on British ships.
The Merchant Shipping (Safety Officers and Reporting of Accidents) (Amendment) Regulations 1984 SI no 93: amend the Safety Officials Regulations 1982 (qv).
The Merchant Shipping (Health and Safety: General Duties) Regulations 1984 SI no 408: impose general duties for the safety and health of persons in British ships.

Norwegian legislation

Petroleum legislation

Legislation applicable directly or indirectly to petroleum activities on the

Norwegian sector of the North Sea continental shelf.

Act of 22 May 1902: The Penal Code: basic criminal law; applies within the realm of Norway; extended to the Norwegian continental shelf by amending Act dated 10 June 1977.

Act of 24 May 1929: Supervision of Electrical Installations: empowers the Norwegian Water Resources and Electricity Board to issue electricity Regulations applicable onshore; also applicable to fixed installations by Regulations dated 3 April 1978 and mobile installations by Regulations dated 11 November 1975.

Act of 16 December 1960: Aviation: regulates flying in Norway; as amended by Act of 10 June 1977; applies to flying within safety zones surrounding offshore installations on the Norwegian continental shelf.

Act of 21 June 1963: Submarine Natural Resources: permits exploitation of natural resources in the sea-bed of Norway under licence; restricted to resources other than petroleum by the Petroleum Act 1985 S67 (qv).

Act of 21 May 1971: Flammable Goods: makes provision for the safe storage, handling and transport of flammable goods.

Act of 14 June 1974: Explosives: makes provision for the safe manufacture, transport and use of explosives.

Act of 4 February 1977: Working Protection and Working Environment Act: makes provision for a safe and healthy working environment, including working hours and employment conditions generally; extended to petroleum activities on the Norwegian continental shelf by Royal Decree dated 1st June 1979.

Act of 22 March 1985: Petroleum Act: licensing and regulation of petroleum activities on the Norwegian continental shelf.

Royal Decrees and Regulations

Of the following Royal Decrees and Regulations, those dated prior to the coming into force of the Petroleum Act 1985 will, unless otherwise stated, continue in force for the time being, but may be progressively re-enacted into the new structure, or remain as Guidelines or standards to be achieved.

Royal Decree of 31 May 1963: assertion of Norwegian sovereignty over the Norwegian continental shelf for the purposes of exploration for and exploitation of natural deposits.

Royal Decree of 25 August 1967: control of foreigners working on drilling platforms on the Norwegian continental shelf.

Royal Decree of 8 December 1972: relates to exploration for and exploitation of petroleum in the sea-bed of the Norwegian continental shelf; repealed with effect from 1 July 1985 by the Petroleum Act 1985 S67 (qv).

Regulations dated 18 April 1973: helicopter decks on platforms engaged in drilling for petroleum on the Norwegian continental shelf.

Royal Decree of 3 October 1975: safe practices in exploration and drilling; repealed with effect from 1 July 1985 by Royal Decree of 28 June 1985, concerning safety in petroleum activities.

Royal Decree of 9 July 1976: safe practices in production; repealed with effect from 1 July 1985 by Royal Decree of 28 June 1985, concerning safety in petroleum activities.

Regulations dated 1 December 1976: for marking of production platforms, etc on the Norwegian continental shelf.

Regulations dated 15 April 1977: for the design of fixed structures on the Norwegian continental shelf.

Royal Decree of 29 April 1977: provision for Safety Delegates and Working Environment Committees under the Working Environment Act (qv).

Regulations dated 25 May 1977: concerning cranes on production installations on the Norwegian continental shelf; applied offshore by Royal Decree of 13 September 1985.

Royal Decree of 3 June 1977: relates to the imposition of coercive fines for enforcement of the Working Environment Act (qv).

Royal Decree of 25 November 1977: provision for hygiene, medical equipment, etc on production installations.

Regulations dated 8 February 1978: regarding life-saving appliances on fixed production installations on the Norwegian continental shelf.

Regulations dated 3 April 1978: for production and auxiliary systems on production installations on the Norwegian continental shelf.

Regulations dated 11 May 1978: concerning state-registered nurses on production installations on the Norwegian continental shelf.

Regulations dated 1 July 1978: provisional Regulations for diving on the Norwegian continental shelf.

Regulations dated 23 October 1978: concerning potable water systems on production installations on the Norwegian continental shelf.

Regulations dated 2 April 1979: for fixed means of access, stairs, ladders, and railings on production platforms on the Norwegian continental shelf.

Regulations dated 2 April 1979: for transfer of personnel to and from production installations.

Regulations dated 2 April 1979: provisional Regulations for living quarters on production installations on the Norwegian continental shelf.

Royal Decree of 1 June 1979: provision for worker protection and safe working environment in connection with exploration for and exploitation of submarine petroleum resources.

Regulations dated 1 August 1980: concerning the medical examination of employees engaged in the production of petroleum on the Norwegian continental shelf.

Regulations dated 9 September 1980: installation and use of radio equipment on mobile installations on the Norwegian continental shelf.

Regulations dated 10 September 1980: for the installation and operation of maritime and aeromobile radio equipment on production installations.

Regulations dated 10 November 1980: approval of survival suits for use in petroleum activities on the Norwegian continental shelf.

Guidelines dated 1 September 1981: for safety evaluation of platform conceptual design.

Regulations dated 23 September 1981: for drilling for petroleum on the Norwegian continental shelf.

Regulations dated 19 March 1982: safety manning in the event of an industrial dispute on the Norwegian continental shelf.

Regulations dated May 1982: concerning physician-in-charge of fixed installations on the Norwegian continental shelf.

Regulations dated 11 October 1982: guidelines for inspection of primary and secondary production structures on the Norwegian continental shelf.

Guidelines dated 10 November 1982: for inspection of primary and secondary structures for production, etc systems.

Royal Decree of 26 November 1982: regulates the labelling and sale of hazardous substances under the Working Environment Act (qv).

Regulations dated 22 February 1983: qualifications for personnel engaged in drilling for petroleum on the Norwegian continental shelf.

Regulations dated 28 December 1983: concerning stand-by vessels in use on the Norwegian continental shelf.

Regulations dated 29 October 1984: for the structural design of loadbearing structures intended for use in the utilisation of petroleum resources.

Royal Decree of 14 June 1985: supplementing Act of 22 March 1985, concerning petroleum activities

Royal Decree of 28 June 1985: Regulations concerning safety in petroleum activities on the Norwegian continental shelf.

Royal Decree of 28 June 1985: Regulations concerning the licensee's internal control in petroleum activities on the Norwegian continental shelf.

Royal Decree of 28 June 1985: supervisory activities regarding safety in petroleum activities on the Norwegian continental shelf.

Supplementary description stipulated by Royal Decree of 28 June 1985: regulatory supervision within the petroleum activities.

Shipping or flag legislation

Legislation applicable to vessels entered in the Norwegian Register of shipping: including mobile installations and other ship-like structures. Applicable world-wide, including while vessels are participating in petroleum activities on the Norwegian continental shelf. Legislation marked * is applied by the Norwegian Petroleum Directorate to Norwegian and foreign flag vessels when engaged in petroleum activity on the Norwegian continental shelf (ie treated as 'petroleum' legislation). Legislation passed prior to the coming into force of the Petroleum Act and its 'Daughter Decrees' (qv) remains in force as 'shipping' legislation; where treated as 'petroleum' legislation it will continue to form part of the standards upon which the new petroleum legislation will be enforced.

Act of 20 July 1893: Maritime Act: provides for the registration of Norwegian ships; as amended by Act of 7 April 1972, includes mobile drilling units.

*Act of 9 June 1903: Seaworthiness Act: provides for the public control of the 'seaworthiness' of ships, including structure, equipment and manning. As amended by the Act of 10 June 1977, extends to Norwegian drilling units; Act as amended also authorises extension to foreign drilling units engaged in petroleum activity on the Norwegian continental shelf.

Act of 30 May 1975: Seamen's Act: terms and conditions, duties and rights of crew on Norwegian ships; was extended to installations by Royal Decree dated 21 October 1976.

Act of 3 June 1977: Working Hours: regulates hours of work of persons employed on Norwegian ships; applied to mobile drilling platforms by Royal Decree of 19 August 1977.

Regulations
*Regulations dated 18 April 1973: helicopter decks on Norwegian and foreign drilling platforms engaged in petroleum activity on the Norwegian continental shelf.
*Regulations dated 15 May 1973: use of radioactive sources on Norwegian or foreign drilling platforms engaged in petroleum activities on the Norwegian continental shelf.
*Regulations dated 10 September 1973 (as amended): for mobile drilling installations, including equipment and structure; applicable to all drilling units engaged in petroleum activities on the Norwegian continental shelf.
*Regulations dated 1 December 1974: electrical installations on board Norwegian and foreign drilling platforms engaged in petroleum activities on the Norwegian continental shelf.
Regulations dated 5 May 1975: construction and operation of Norwegian mobile drilling units.
Regulations dated 16 June 1975: mustering on board Norwegian ships, including mobile drilling platforms.
*Regulations dated 29 August 1975: drilling Regulations applicable to all mobile drilling platforms operating on the Norwegian continental shelf (repealed, except for Regs 6 and 9).
Regulations dated 22 March 1976: construction and operation of passenger and cargo lifts on Norwegian ships and mobile drilling platforms.
Regulations dated 15 November 1976: Protection Supervisors and Protection and Environment Committees on Norwegian ships and mobile drilling vessels and platforms.
Regulations dated 21 March 1977: welding equipment on Norwegian ships and mobile drilling platforms.
Regulations dated 19 August 1977: working hours on Norwegian mobile drilling platforms.
*Regulations dated 31 January 1978: deck cranes on Norwegian and foreign drilling units engaged in petroleum activities on the Norwegian continental shelf.
*Regulations dated 31 January 1978: arrangements and safety measures on Norwegian and foreign drilling units engaged in petroleum activity on the Norwegian continental shelf.
*Regulations dated 1 July 1978: provisional Regulations for diving on the Norwegian continental shelf; applicable to Norwegian or foreign units.
Regulations dated 8 September 1980: telecommunications on board mobile drilling units and accommodation vessels on the Norwegian continental shelf.
Regulations dated 30 March 1981: medical examinations for employees on Norwegian ships, including drilling platforms.
Regulations dated 26 June 1981: working hours for diving personnel on board Norwegian ships and drilling platforms.
*Regulations dated 23 September 1981: drilling Regulations applicable to all drilling

units engaged on the Norwegian continental shelf.

Regulations dated 11 December 1981: certification of personnel on Norwegian drilling units.

*Regulations dated 3 February 1982: life-saving appliances on Norwegian mobile installations; also foreign mobile installations engaged in petroleum activity on the Norwegian continental shelf.

Regulations dated 23 March 1982: manning requirements for personnel on Norwegian drilling units and other mobile installations.

Regulations dated 23 March 1982: qualification requirements for personnel on Norwegian drilling units and other mobile installations.

*Regulations dated 11 June 1982: construction and equipment of living quarters on Norwegian mobile installations; applicable also to foreign drilling units engaged in petroleum activities on the Norwegian continental shelf.

Regulations dated 31 January 1984: fire safety measures on Norwegian drilling units and other mobile installations.

Regulations dated 10 April 1984: control of diving systems on Norwegian offshore units.

Regulations dated 7 January 1985: potable water systems on Norwegian mobile drilling units.

*Royal Decree of 28 June 1985: licensee's internal control in petroleum activities on the Norwegian continental shelf; particularly the control of both Norwegian and foreign mobile drilling units.

*Royal Decree dated 28 June 1985: concerning safety in petroleum activities on the Norwegian continental shelf; particularly supervisory activities in relation to mobile drilling units of both Norwegian and foreign registry.

Regulations dated 25 April 1986: drilling installations and equipment on Norwegian mobile drilling units.

Regulations dated 13 January 1986: safety arrangements above and below deck on Norwegian drilling units and other mobile units.

Regulations dated 13 January 1986: use of radioactive sources on Norwegian drilling platforms and other mobile units.

Regulations dated 13 January 1986: construction and operation of Norwegian drilling platforms and other mobile offshore units.

Regulations dated 13 January 1986: deck cranes on Norwegian mobile units.

Regulations dated 13 January 1986: approval of survival suits in offshore petroleum activities.

Regulations dated 13 January 1986: helicopter decks on Norwegian drilling units and other mobile offshore units.

Regulations dated 13 January 1986: installation and use of maritime and aeromobile radio equipment on Norwegian drilling units and other mobile units.

Regulations dated 13 January 1986: construction and equipment of living quarters on Norwegian drilling units and other mobile offshore units.

Regulations dated 13th January 1986: life-saving appliances on Norwegian drilling units and other mobile offshore units.

Regulations dated 28 June 1985: owner's duty with regard to internal control of Norwegian mobile offshore units.

Regulations dated 7 January 1985: potable water system and supply on Norwegian mobile drilling units and other mobile offshore units.

Other legislation

Either shelf or flag legislation applicable to offshore petroleum safety.

United States

The Outer Continental Shelf Lands Acts 1953 et seq: assert US federal governmental claims to petroleum resources under these lands; also make provision for health and safety of employees engaged therein. Enforcement by the US Minerals Management service and by the US coast guard, which is the 'lead Federal Agency' for this purpose.

American Shipping Acts: empowers the US coast guard to make Regulations for, *inter alia*, the safety of vessels registered in the American Register of Shipping; this power extends to mobile drilling units.

Occupational Safety and Health Act 1970: federal onshore safety legislation with specific offshore application; enforced offshore by US coast guard under a MOU with Occupational Safety and Health Administration.

Regulations issued January 1979 (46 CFR Chapter 1 -A): inspection and certification, design and equipment for mobile drilling units registered under the US flag. These Regulations embody the Safety of Life at Sea and the Load Line Conventions (qv). Enforced by the US coast guard.

Regulations issued April 1982 (46 CFR Chapter 1-A): require foreign mobile drilling units operating on the outer continental shelf to meet comparable standards to those met by US units.

Regulations issued 4 March 1982 (33 CFR Chapter N): for outer continental shelf lands activities relate to cranes and personnel matters on production platforms; enforced by Minerals Management Service.

Canada

Canada Shipping Acts: regulates vessels registered in Canada, also vessels within Canadian waters.

Canadian Oil and Gas Drilling Regulations 1980: regulate drilling on the Canadian continental shelf.

Newfoundland and Labrador Petroleum Drilling Regulations 1977: regulate drilling on that part of the continental shelf claimed by Newfoundland.

Australia

Petroleum (Submerged Lands) Act 1967: (as amended by the Seas and Submerged Lands Act 1973) asserts Australian federal jurisdiction over the petroleum resources of the Australian continental shelf, including power to license operations, which may extend to the imposition of safety obligations upon the licensee.

Petroleum (Submerged Lands) (Amendment) Act 1980: provides for the creation of joint federal-state regulatory authorities for the offshore industry.

Navigation Act 1912: provides for the registration of Australian ships, and empowers the federal government to regulate standards on these ships; has been applied to Australian drilling vessels, also to vessels of any nationality engaged in petroleum activity on the Australian continental shelf. Once a unit begins to drill, it comes under the Petroleum (Submerged Lands) Acts and Regulations.

Marine Order (Part 47): detailed requirements relating to the design and operation of mobile offshore drilling units; issued by the Department of Transport.

Appendix 2

Reports and other publications relating to safety offshore, in order of publication.

Code of Safe Practice for Drilling and Production in Marine Areas. Institute of Petroleum 1964.

Inquiry into the causes of the accident to the drilling rig Sea Gem Ministry of Power Report, Cmnd 3409, HMSO, London, 1969.

Report of Committee on Safety and Health at Work (Robens Report), Cmnd 5034, HMSO, London, 1972.

Development of the Oil and Gas Resources of the United Kingdom Department of Energy Annual Reports 1973-1985, London.

Interdepartmental Committee on Marine Safety, Report 1976-77.

Report No 65 to the Storting (1977–78): The uncontrolled blow-out in the Ekofisk Field on 22 April 1977 Ministry of Local Government and Labour, Oslo, Norway.

Safety of Seamen at Work: Report of the Steering Committee on the Safety of Merchant Seamen at Work Department of Trade, HMSO, London, 1978.

Sikorsky S.61N Helicopter G-BBHN Report of the accident in the North Sea East of Aberdeen on 10 October 1977, Department of Trade, HMSO, 1978.

The Fire in HMS Glasgow, 23 September 1976 Health and Safety Executive, HMSO, 1978.

Assessment of the suitability of stand-by vessels attending offshore installations Instructions for the guidance of surveyors, Department of Trade, London, 1978.

Offshore Safety Report of the Burgoyne Committee 1980, HMSO, Cmnd 7841, London.

Occupational Health and Safety in the North Sea: Report to the Health and Safety Executive Barrett BN and Howells RWL 1980.

Safety Problems in the Offshore Petroleum Industry International Labour Office, Geneva, Switzerland, 1980.

Working Conditions and Working Environment in the Petroleum Industry including Offshore Activites International Labour Office, Geneva, Switzerland, 1980.

Model Code of Safe Practice for the Petroleum Industry: Part VIII – Drilling, Production and Pipe-line Operations in Marine Areas Institute of Petroleum, Heydon and Co, London, 1980.

Health and Safety Guide for Oil and Gas Well Drilling and Servicing National Institute of Occupational Safety and Health, Cincinatti, USA, 1980.

Safety and Offshore Oil: Report of the Committee on Assessment of Safety of Outer Continental Shelf Activities Washington, 1981.

Safety and Health in the Construction of Fixed Offshore Installations in the Petroleum Industry. International Labour Office, Geneva, Switzerland, 1981.

Fires and Explosions on Fixed Platforms in the Norwegian Sea Det Norsk Veritas, Oslo, Norway, 1981.

Fires and Explosions on Offshore Platforms: A Statistical Survey of Gulph of Mexico Data Det Norsk Veritas, Oslo, Norway, 1981.

Report No 67 to the Storting (1981-82): The Alexander L Kielland Accident Ministry of Local Government and Labour, Oslo, Norway.

Commission of the European Communities: Proceedings of International Symposium on Safety and Health in the Oil and Gas Extractive Industries Jean Monnet, Luxembourg, 19–20 April 1983.

11th World Petroleum Congress, 1983 Report 1984, Wiley and Sons, London.

Royal Commission on the Ocean Ranger Marine Disaster: Report Canada August 1984.

Other references

Anon Coastguard pre-emption of The Occupational Safety and Health Act on the Offshore Continental Shelf. *5 Maritime Law 288 (1980).*

Barrett, B and Howells, RWL Occupational health and safety in the North Sea. *131 NLJ 3254 (1981).*

Barrett, B and Howells, RWL Safe systems for exploiting the petroleum resources of the North Sea, *33 ICLQ 811 (1984).*

Bell, J Death of an oil platform. *New Scientist* (27 February 1986).

Bentham, RW Role of law in offshore activities. *109 Ocean Management (1978).*

Birnie, P The legal background to North Sea oil and gas development in: *The Political Implications of North Sea Oil and Gas* Eds. M Saeter and I Smart (1975).

Birnie, P Did failures in the North Sea legal regime contribute to the Ekofisk blow-out? *119 Ocean Management (1978).*

Bull, HJ Legal and administrative framework for Norwegian offshore petroleum activities. *10 International Business Law (1982).*

Carson, WG The Other Price of Britain's Oil Martin Robertson and Co (1982).

Chambers, I Application of International Labor Conventions offshore; an approach. *50 International Lab Rev 120; 395/409 (1981).*

Churchill, RR Law and offshore oil development; the North Sea Experience *29 ICLQ 173 (1980).*

Daintith, T and Willoughby, GDM United Kingdom Oil and Gas Law. Oyez (1977)

Daintith, T and Willoughby, G Manual of Oil and Gas Law. Sweet & Maxwell (1984).

Daintith, T and Willoughby, G UK Oil and Gas Law (looseleaf) Sweet & Maxwell (1984)

de Mestral, ALC Law applicable to the Canadian East Coast Offshore, *21 Alta Law Rev 63 (1983).*

Donelan, EJ Offshore exploration and exploitation (United Kingdom and Ireland) *133 NLJ* 461 (1983).

Evans, AC North Sea oil and EEC law *16 International Lawyer* 529. (1982).

Goldsworthy, P Ownership of the Territorial Sea and Continental Shelf of Australia *50 Australian LJ* 175 (1986).

Hayashi, T Offshore Casualties in Canadian waters *21 Alta Law Rev* 165 (1983).

Herman, LL Need for a Canadian Submerged Lands Act *58 Can Bar Rev* 518 (1980).

Howells, RWL and Barrett, B The Health and Safety at Work Act (2nd ed) IPM Press (1982).

Inions, NJ Newfoundland offshore claims *19 Alta Law Rev* 461 (1981).

Jones, WSL Personal injury – Offshore Oil Operations *Natural Resources Lawyer* 681 (1972).

Kitchen, Jonathan Labour Law and Offshore Oil. Croom Helm (1977).

Kjanstad, A Conference paper on Norwegian accident claims. Caracas (1985).

Mankabady, S Ocean oil and gas drilling and the law *30 ICLQ* 267 (1981).

McEvoy, J Atlantic Canada – constitutional offshore regime *Dalhousie LJ* 284 (1984).

Millard, WH Legal environment of the British oil industry *18 Tulsa L J* 384 (1983).

Morris, JW The North Sea Continental Shelf: Oil and Gas Legal Problems *2 International Lawyer* 191 (1968).

Okere, BO Techniques of international maritime legislation *30 ICLQ* 512 (1981).

Selmer, KS Interactions between insurance and tort theories in the Norwegian law of Personal Injuries *18 Amer Jo of Comp Law* 145 (1970).

Thomas, M and Steel, D Temperley's Merchant Shipping Acts (7th ed) Stevens (1976).

Thompson, AG New offshore petroleum legislation; (Petroleum submerged Lands) Amendment Act 1980 *1 Australian Bus L Rev* 226 (1983).

Whitehead, H United Kingdom Offshore Legislation guide (looseleaf) (1985).

Wright, C Routine deaths: fatal accidents in the oil industry *Sociological Review* 265 (1986).

Appendix 3

Accidents, disasters and dangerous occurrences

1968: loss of drilling rig *Sea Gem* off Humber, British sector of the North Sea; 13 fatalities.

1968: helicopter crash on *Barracouta Platform*, Bass Strait, Australia; 3 fatalities.

1971: major blow-out and fire on *Marlin Platform*, Bass Strait, Australia; no casualties.

April 1977: uncontrolled blow-out in the *Ekofisk Field* in the Norwegian sector of the North Sea; no fatalities.

1978: running aground off Guernsey of drilling rig *Orion*; no fatalities.

January 1979: sinking of new drilling rig off Hartlepool; no fatalities.

March 1979: sinking of drilling rig off North Wales; no fatalities.

May 1979: Drilling rig collapses off Galveston, Texas; 8 fatalities.

August 1979: loss of diving bell in *Thistle Field*, British sector of North Sea; 2 fatalities.

November 1979: drilling rig collapses off China; 72 fatalities.

December 1979: crane barge adrift in *Tartan/Frigg Fields*; 500 evacuated, but no fatalities.

January 1980: drilling rig collapses in Gulf of Mexico; no fatalities.

March 1980: gas leak on drilling platform in *Argyll Field*, British sector of North Sea; 7 fatalities.

March 1980: explosion and fire on platform in Gulf of Mexico; 7 fatalities.

March 1980: overturning of accommodation semi-submersible *Alexander L Kielland* in the Ekofisk area of the Norwegian sector of the North Sea; 123 fatalities.

April 1980: *Henrik Ibsen*, sister ship to the Alexander L Kielland, develops list off Stavanger and evacuated; no fatalities.

August 1980: rig overturned by hurricane off Louisiana; 4 fatalities.

August 1980: tanker collides with foundations of new platform in Mexican Gulf, catches fire; no fatalities.

October 1980: gas cloud envelops platform off Saudi Arabia; 19 fatalities.

January 1981: blow-out in the Mexican Gulf; no fatalities.

February 1981: floating rig burns out of control in Mexican Gulf; no fatalities.

June 1981: rig turns over and sinks off Angola; no fatalities.

August 1981: Wessex helicopter accident in British sector of North Sea; 13 fatalities.

October 1981: BP rig adrift off Ireland; no fatalities.

November 1981: two drilling rigs break anchor chains and drift in British sector of North Sea; no fatalities.

December 1981: cracks discovered in structure of semi-submersible drilling platform in North Sea; 400 evacuated, but no fatalities.

February 1981: loss of semi-submersible *Ocean Ranger* off St John's Newfoundland; 84 fatalities.

1983: capsize of drilling rig *Glomar Java Sea* in storm off China; 81 fatalities.

September 1983: loss of jack-up rig *Key Biscayne* in Timor Sea, off Australia; no fatalities.

October 1985: gas blow out on drilling rig *West Vanguard*; 1 fatality.

November 1986: helicopter accident in British sector of North Sea; 45 fatalities.

Index